T0320139

Nanotechnology for a Sustainable World

Nanotechnology for a Sustainable World

Global Artificial Photosynthesis as
Nanotechnology's Moral Culmination

Thomas Faunce

*Professor, Australian National University, Australian Research
Council Future Fellow*

Edward Elgar

Cheltenham, UK • Northampton, MA, USA

Published by
Edward Elgar Publishing Limited
The Lypiatts
15 Lansdown Road
Cheltenham
Glos GL50 2JA
UK

Edward Elgar Publishing, Inc.
William Pratt House
9 Dewey Court
Northampton
Massachusetts 01060
USA

A catalogue record for this book is available from the British Library

Library of Congress Control Number: 2011942538

ISBN 978 1 84844 671 7

Typeset by Columns Design XML Ltd, Reading
Printed and bound by MPG Books Group, UK

Contents

Acknowledgements

The author acknowledges financial support provided by the Australian Research Council through a Future Fellowship. He also thanks the Brocher Foundation in Geneva (http://www.brocher.ch/) for a visiting scholarship in 2009 and for a Brocher alumni meeting in 2011.

I'm grateful to the scholars who commented on early drafts of the work including Peter Zweifel for discussions about private insurance, Leslie Zines for comments about the social contract, Sidney Paulson for sharing insights about Hans Kelsen and the grundnorm. For opportunities and discussions in relation to artificial photosynthesis appreciation must particularly be expressed to Peidong Yang, Dan Nocera, Gary Brudvig, David Tiede, Chris Moser, Craig Hill, Stenbjörn Styring, Ann Magnuson, Johannes Messinger, Graham Fleming, Bill Rutherford, Philippe Schild, Wolfgang Lubitz, Holger Dau, Paul Mulvaney, Doug MacFarlane, Michael Graetzel, Jan Anderson, Ron Pace, Warwick Hillier, Jim Barber, Elmars Krausz, Eva-Mari Aro, Fred Chow and Graham Farquhar. John White generously included me in his research team for light scattering and neutron scattering experiments with nanoparticles, providing insights into the world of physics I would never otherwise have experienced.

Particular thanks should go to Claire Findlay in the federal Department of Industry, Innovation, Science and Research and Christine Debono, as well as others at the ANU College of Law, UNESCO, the scientists from research labs across the world and the citizens of Lord Howe Island, who facilitated the 'Towards Global Artificial Photosynthesis: Energy, Nanochemistry and Governance' conference at that World Heritage location in August 2011.

Preface

Don't ever forget these things:
The nature of the world.
My nature.
How I relate to the world.
What proportion of it I make up.
That you are part of nature, and no one can prevent you
from speaking and acting in harmony with it, always.
– Marcus Aurelius, *Meditations*

Because of the speed with which events and scientific breakthroughs are communicated and turned into marketable products in the contemporary world, many books dealing with emerging technology are virtually out of date at the time of publication. This book seeks to avoid that disadvantage by setting its arguments against what will hopefully remain timeless backgrounds: the natural beauty of our Earth and the human struggle to not only discern the laws of the physical universe, but also to cohere with them by actions that promote great social virtues such as justice and equity. It makes a case, in that context, for how the most promising new technology of our age (nanotechnology), when properly governed, is likely to help resolve the looming social and environmental crises that threaten our existence.

These pages also chart a personal intellectual and moral journey in which the natural environment has been an important co-conspirator. They plot thoughts and dreams about environmental law and justice that began emerging while I was working with Justice Lionel Murphy as a young lawyer by Lake Burley Griffin in Canberra. They take in others about environmental sustainability that arose by Henry Thoreau's pond near the small town of Concord in Massachusetts and amongst a 'New Age' community in a pine forest by the village of Findhorn in Scotland. They incorporate views about public health arising from experiences practising emergency medicine in the Wagga Wagga general hospital by the Murrumbidgee River and the Alfred Hospital Intensive Care Unit near the Yarra River in Melbourne. They draw upon reflections on human purpose that arose by the ruins of the ancient Buddhist universities of Nalanda and Vikramasila in northern India, as well as a Buddhist monastery (Wat Hin Maak Peng) next to the Mekong River in northern Thailand. They reflect experiences fighting (in the media and in Senate committees) supranational corporations attempting to use trade agreements to alter democratically endorsed health policy.

What follows is a personal opinion on what should be nanotechnology's pre-eminent role in sustaining our world. Those particular ideas developed whilst on a Brocher Foundation Fellowship in Geneva, while participating in neutron scattering experiments at Lucas Heights in Sydney and Grenoble in France as well as, most particularly, through discussions with experts in artificial photosynthesis at physics and chemistry conferences – beside a melting glacier at Obergurgl in Austria and amongst the World Heritage listed beauty of Lord Howe Island.

This small book is about how humanity should best use its newest building blocks, material and conceptual – both in technology and in governance. It involves a consideration of some of the most fundamental laws of our world – equations relating, for example, to time, space and consciousness, as well as moral injunctions emerging from our foundational social virtues and collective conscience. It considers how important it is to promote the understanding that reality in not only its physical but its moral and jurisprudential forms is shaped by laws of a geometrical structure only partially revealed to us by our senses and common sense. It examines the hypothesis that worldwide use of nanotechnology should take place in a normative context coherent with its foundational science. A major outcome of such reasoning may be to emphasise environmental sustainability as a foundational social virtue in global governance theory and structures.

The idea that social laws should draw their fundamental impetus and insights from science is not by any means the dominant view amongst contemporary policy or law-makers. Many continue to see trade-offs between short-term political expediency, private corporate interests and religious ideology (in one of its numerous, institutionally conflicting forms) as the inevitable source of governance norms at local, national and global levels. The debate over global legal responses to scientifically proven anthropogenic climate change is a pertinent example of such conceptual positions. Altruistic and beneficent goals are far from inevitable components of any new technology's widespread use; indeed it seems to follow some implicit societal principle that a large proportion of people and political parties deny such aims are presently or should ever be part of everyday human experience.

History has proven how impossible it is to accurately predict technological change. Indeed, from a historical perspective any technology may be viewed as but a hardening of consciousness, a temporary solidification of the onward thrust of the human impulse, as shaped by material and moral pressures. This text aims to stimulate researchers, students and policy-makers to become more critically interested in how nanotechnology, properly coordinated on a global scale, can assist in resolving some of the most urgent problems facing humanity and our planet. At a more fundamental level it is designed to encourage such people to think better, if only at first for the simple joy of doing so, about how nanoscience may help us to develop the legal norms and governance structures that will allow nanotechnology to play this valuable role.

Let's now turn to Chapter 1 and its introduction to some basic concepts underpinning nanoscience and environmental sustainability in governance. That chapter will end by questioning whether humanity needs a Big Science nanotechnology for a sustainability project focused on problems such as expanding human energy and food demands, poverty and environmental degradation. It will consider, if such a project can be supported ideally, by what practical criteria we should select its subject and best construct the grand design of its governance architecture. The following chapters set out the major obstacles to that vision, the possible candidates for such a project and a final explication of the most feasible amongst them.

1. Introduction

There is also a rhythm and a pattern between the phenomena of nature
which is not apparent to the eye, but only to the eye of analysis;
and it is these rhythms and patterns which we call Physical Laws
– Richard Feynman, *The Character of Physical Law*

It is impossible to conceive anything at all in the world, or even out of it,
which cannot be taken as good without qualification, except a *good will...*
to make itself a universal law is itself the sole law
which the will of every rational being spontaneously imposes on itself
without basing it on any impulsion or interest.
– Immanuel Kant, *Groundwork of the Metaphysic of Morals*

1.1 NANOTECHNOLOGY, VIRTUES AND LAWS FOR A SUSTAINABLE WORLD

During my last years in law school at the Australian National University
my interests gravitated towards constitutional, international, science and
environmental law, as well as haiku poetry (through a Japanese language
subject in my arts degree). I was interested in law as an ideal. When required
to study the compulsory subject property law I achieved a certain notoriety
amongst the academic staff with an essay entitled 'midline cerebral com-
missurotomy and the law of mortgages', arguing for a historical progression from right-brain symbolic ceremonies of livery of seisin to the left-
brain written documents that today govern sale and purchase of land.

It was fortunate that my first job after graduating was as research
associate to Justice Lionel Murphy in the High Court of Australia, situated
beside Lake Burley Griffin in Canberra. Justice Murphy had achieved
honours in both science and law at university and possessed a mind that not
only sparked ideas, but as an Attorney General in the federal government
was also courageous in applying them to further the legislative pursuit of
justice and equity. In his office, where other judges had a picture of the
Queen, Justice Murphy placed an image of a nitrogen-abundant supernova
remnant (SNR) N86 in the Large Magellanic Cloud. This had been named
after him by astronomers at the Australian National University's Mount
Stromlo Observatory. A supernova remnant results from a star exploding

its atoms outward at velocities near 1 per cent the speed of light and temperatures over 10 million K. The analogy those astronomers perceived with Justice Murphy's impact on the law remains pertinent. The laws he created (both as Attorney General and as a judge) continue to illuminate the firmament of individual civil and political human rights against governmental injustice and inequity.

From the judge's chambers you could see beyond the lake sunlight and clouds play above the beautiful mountains surrounding Australia's well-treed national capital. It was an appropriate setting for one of the most important cases the High Court heard that year. This was the constitutional challenge to federal legislation preventing the construction of a hydro-electric dam on the last wild river in Australia – the Franklin River in Tasmania.

The appeal papers proved the dam was a wonderful piece of technology. It promised to increase the renewable energy available to significantly provide for Tasmania's industrial and domestic use of electricity. Yet the dam would also have destroyed a beautiful wilderness, as well as archeologically significant sacred sites of the aboriginal people of Tasmania. The case arose from a nationwide campaign of protest and civil disobedience that finally resulted in the election of a federal government prepared to pass legislation preventing the dam. The appeal against that legislation was argued chiefly over interpretations of the power of the federal government to pass laws about corporations and foreign affairs (the river had been placed on the UNESCO World Heritage List that created binding international law obligations for its protection for signatory nations such as Australia).

The Franklin River was saved – by a narrow judicial majority (4:3). It flows free today, from source to sea, in a National Park. A decade later, after graduating from medical school, I canoed down that river and experienced how serenity and idealism seem to permeate your being from its remarkable natural beauty. Whatever might have been the technical legal grounds on which the Franklin River Dam case was fought and decided, those of us involved closely in that struggle felt that here was an instance of law responding to a new call in social conscience – not necessarily to right injustice or inequity, but to prevent environmental degradation in accordance with international obligations. Involvement in that case started me thinking about how law (and international law in particular) should shape our use of new technologies so that they benefit rather than degrade our environment. It also emphasised to me how a society, by manifesting a commitment to apply a universally applicable principle in the face of obstacles, can shape virtues, a collective character.

Nanotechnology fits within a category of historical examples of transformative technologies including tool and shelter making, growing and cooking food, the wheel, boats, metallurgy, hay-making, knitting, spinning machines, paper, the printing press, steam power, internal combustion engines, aircraft, electricity, nuclear energy, satellites, spacecraft, the Internet and genomics. Like those examples, nanotechnology is about to set the symbolic tone and practical agenda for a new stage of human moral as well as material progress.

So what is nanotechnology? Let's first examine some definitions. A nanometre, or 10^{-9} m, is a billionth of a metre. An atom is about one cubic nanometre (the important unit of one Ångström is approximately equal to the width of an atom and is 0.1 nanometre). Nanotechnology, as commonly specified, involves research in physics and chemistry and development of products utilising engineered ultra-small particles (with unusual names like quantum dots, oxides, nanocomposites, nanowires, Fullerenes and single or multiwalled carbon nanotubes) having at least one dimension less than approximately 100 nm and related distinctive properties. Yet, this definition of nanotechnology (more precise ones are being developed by the International Organization for Standardization (ISO)) seems sanitised from the great global challenges this new research area should be responding to.

The wavelengths of the form of electromagnetic radiation we know as visible light fall roughly into the range 400–700 nm. This means that atoms, as well as the nanoparticles made from them, cannot be viewed using an optical microscope, but can be discerned through techniques such as the scanning electron microscope, as well as light, X-ray and neutron scattering. Below the nanoscale is the subatomic picoscale (a picometre is a trillionth of a metre), the realm of quantum physics and the smallest scale of matter, space and time – the Planck scale. After that come realms of metaphysics where matter, time and space may merge into a single field of 'mental-type stuff'. As James Jeans put it in his marvellous book *The Mysterious Universe*: 'the rolling contact of our consciousness with the empty soap-bubble we call space-time […] reduces merely to a contact between mind and a creation of mind […] the tendency to thinking the way which, for want of a better word, we describe as mathematical'.

Part of the reason for the considerable contemporary industrial and scientific interest in nanotechnology is that the physical and chemical properties of engineered nanoparticles (ENPs) differ from bulk equivalents in potentially temporally very useful ways. ENPs, for example, have larger surface area per unit mass (increasing strength and binding but also biological reactivity) and many have quantum physical effects below about

10 nm involving altered electrical conductivity, catalytic properties, wavelength of emitted light and magnetisation. The benefits of things engineered from nanoscale components include proportionally greater surface area and strength with less weight as well as enhanced energy storage and transmission (particularly from their capacity to harness quantum effects).

Nanotechnology offers to fill opportunity niches in a wide range of manufacturing, food, health and energy systems. Vast amounts of money are invested in nanotechnology research globally. Many of its most marketed applications appear to reflect our immaturity as a species (given the public health and environmental challenges we face) – sunscreens that look nice, lighter and stronger golf clubs (presumably so as not to overload the golf cart and resist being smashed against trees), socks and shirts that don't smell, washing machines without grime.

Some of the most common marketed nanomaterials include carbon black used in products such as automobile tyres and antistatic textiles as well as to colour rubber, ink and leather. Also in widespread use are carbon nanotubes (nanometre scale rolled sheets of graphite). Their strength has seen them applied to consumer items like tennis rackets, golf clubs, skis and bicycles, as well as building materials including pavers. Owing to their large surface area, high electrical conductivity and adsorption capacity, carbon nanotubes have applications in energy storage products, electromagnetic shielding and super capacitors. There are major safety concerns about inhalation of carbon nanotubes – their relative length and biopersistence creating a heightened risk (based on animal models) of asbestosis-like lung injury.

Silicon-based nanomaterials are utilised in biosensing and bioimaging, batteries, microelectronics, and in photovoltaics. Nanosilver is widely used in consumer products (food packaging, washing machines, socks and shirts) as an antimicrobial. Major issues surround the accumulation of nanosilver in our sewerage treatment systems, waterways and the food chain. Nanotitanium dioxide and nanozinc oxide provide UV cosmetically attractive protection in 'invisible' sunscreens and face creams as well as outdoor paints. Although cell models confirm damage from these nanoparticles, the dead outer layer of human skin (stratum corneum) prevents most of them getting into the blood stream, though this provides less comfort for those with damaged or frequently cut skin.

Nanocerium oxide is commonly employed in semiconductors, in electrolytes for solid oxide fuel cells, in oxygen sensors and in car exhaust catalysts. It has been developed as a product that reduces fuel consumption and greenhouse gas emissions (CO_2), and particulates emissions when added to diesel fuel.

This is just a small sample of the current uses of nanotechnology. Nanotechnology also offers the promise of revolutionising fields as diverse as building materials and clothing, energy generation, food production, water and soil purification, medicines, computing and weapons production, as well as the process of manufacture itself.

It is quite realistic to imagine a future where almost every product we use has a nanotechnology component, where nanotechnology has become a ubiquitous part of our civilisation, integrating itself not only into our social structures, but also our bodies and the very way we exercise our freedom of will and conscience. This book tries to embrace such a vision – of a nanotechnology-embedded world – then to imagine its moral and legal implications. Inevitably this involves critically analysing the jurisprudential foundations and intersections of different realms of law with new technology – something I expect Justice Lionel Murphy would have found fascinating.

Most official definitions of, or business plans to develop nanotechnology globally lack a moral or normative element – a developed connection to the great ethical principles, societal virtues and legal rights that have shaped human civilisation and that many of the most influential human thought-leaders have posited are part of the natural structure of the world.

In fact, as we'll examine in subsequent chapters, many of the major stakeholders involved in researching and developing nanotechnology are driven by their constitutive documents to self-interestedly prioritise maximisation of shareholder profits (in the case of supranational corporations) and equally self-interestedly defined sovereign interests (in the case of nation states). Indeed, the dominant governance focus on nanotechnology has been on whether its use in consumer products presents us with unusual and important toxicological problems. Some civil society organisations (such as Friends of the Earth) even claim there should be a moratorium on the use of ENPs until such toxicological issues are thoroughly resolved.

Having accepted the imminent existence of a nanotechnology-based global society, this book then considers a related postulate – whether such a society is more likely to explore the idea that social laws may be developing in synergy with (and in fact might represent harmonics of) the physical laws underpinning the universe even as they are shown not to readily correlate with our common experience.

It is now well established in modern physics, for example, that many aspects of reality are true, despite conflicting with how we generally reason about the world based on sensory information. Some notable examples include light's uniform speed regardless of its source, the extinction of matter in black holes, the invisibility of dark matter, the slowing down of time (relative to us) for an object whose speed approaches that of light, the capacity of matter to warp time and space, the alteration of particle position through the act of observation and the simultaneous existence of all matter and energy (presumably including us) as a particle and wave.

The world we are bequeathing future generations (where nanotechnology has become ubiquitous) may be one in which scientists verifying key aspects of string theory with equipment such as the Large Hadron Collider have unified the mathematical equations (laws) of microscale quantum mechanics and macroscale general relativity. They might, for example, have proven the existence of 'supersymmetric' partners for every known subatomic particle species, evanescent mini-black holes created by collisions of subatomic particles, or the existence of more than three dimensions of space and one of time that dilute gravitational force and the energy from particle collisions over tiny distances; indeed, that we in fact reside in a multiverse.

One hypothesis developed here is that such a world may also be one in which the fundamental symmetry increasingly revealed by geometry, mathematics and physics has assisted the development of social virtues and laws more coherent with such understandings, for example respecting our environment and promoting the expansion of human consciousness to identify with a vantage point broader and more timeless than our own. It will be argued here that widespread use of nanotechnology actually may promote this global normative evolution ('norm' in this context meaning social principle or rule) and become a crucial bridge to ages predominantly influenced by even more subtle forms of technology and hence more refined moral conceptions.

As mentioned, influential and well-respected non-governmental organisations are robustly opposed to the widespread use of nanotechnology in large part because they allege that process is failing to follow principles and rules that adequately limit the harm it may do humans and the environment. Part of the case advanced here is that such a restrictive governance approach, whilst commendably focused on safeguarding public health and the environment, fails to take account of the critical role that nanotechnology can play in advancing our moral progress and (which may turn out to be the same thing) coherence with the fundamental symmetry and harmony revealed by physical laws.

Let's approach this insight in another way. The physicist Richard Feynman gave the spur to the nanotechnology revolution with a 1959 lecture claiming that there was 'plenty of room' to build things at the scale of less than 100 nm (10^{-9} m). Feynman once said that true discovery (such as that likely to lead to nanotechnology) comes with an accumulation of paradoxes, with existing laws being proven to give inconsistent results. Feynman believed (if one can say that in connection with such an iconoclastic experimentalist) that we sought new physical laws to reduce complexity; we think of an ideal like symmetry, put the information in mathematical form and then guess the equations. Feynman influentially recognised that our world tends to reflect geometric and mathematical patterns, but thought it unlikely that the laws of geometry extend down into infinitely small space.

The task of proving that 'laws' go all the way down in reality is complex because as matter and time reaches its smallest extent (the Planck scale) quantum effects make everything including points, velocity, length and curvature fluctuate, become indeterminate and uncertain as soon as we try to observe them. Thus at a Planck length of 1.6161×10^{-35} (which squared gives the Planck area of 2.61177×10^{-70} m^2), or a Planck time of 5.39072×10^{-44} s, or Planck mass of 2.17665×10^{-8} kg, gravity starts to become as strong as the electromagnetic force and, with the strong and weak nuclear forces, distorts the very essence of space and time. Planck units, like the elementary electric charge and speed of light, are natural 'laws' in that they are based on universal physical constants, rather than any historically located human definition.

The geometer Shing-Tung Yau considers that geometry might be the ultimate source of such 'natural laws', shaping multiple dimensions of space and time even in the absence of matter. The string theorist Edward Witten has argued for a new type of quantum geometry in which all the basic 'laws' of matter might be harmonics of strings vibrating (and dissipating gravity and energy) in those multiple additional dimensions.

The notion explored here is that embedding nanotechnology in our world is likely to take the reasoning behind its governance arrangements towards such speculations if only initially as a source of analogy. This book, in other words, considers whether the 'laws' of global governance systems then may come to be viewed at their most fundamental as not merely contingent manifestations of political compromise and judicial interpretation, but (when considered from the appropriate perspective) likewise aspects of a universal geometric and mathematical symmetry.

Such a potential synergy of basic 'laws' of nanotechnology and foundational principles of global governance is illustrated in the life and work of the futurist Buckminster Fuller. Fuller believed that the fundamental geometry of the physical universe involved tetrahedrons (a perspective very similar to that of esteemed ancient philosophers Plato and Pythagoras). He drew on this insight to create the stable geodesic domes that made him famous. Graphite nanostructures shaped like a geodesic dome (with applications in superconductivity) were named Fullerenes in recognition of Fuller.

Yet Fuller's philosophy of the basic building blocks of the universe also stimulated his idea that human society had to renegotiate its basic principles of social interaction. We were no longer, Fuller thought, living in a time where we needed to compete violently for resources. Instead, our survival and flourishing, when viewed as a system trying to operate most efficiently, was critically dependent on our ability to conserve and recycle.

Another way of expressing this is to suggest that it is the virtues of curiosity and love of harmony, as well as more unpredictably valuable character traits such as pride and determination, that have assisted our species' progressive understanding not only of natural 'laws' such as those underpinning physics, biology and chemistry, but also synergistically of the closely related principles, rules and laws designed to govern our relationships with each other and the environment. On such a view the human species (at least in its most intellectually refined manifestations) seems to love to champion geometrically elegant ideas and that has given it a practical survival advantage in terms not only of technological development, but also of social governance. Love and conscience may themselves be manifestations of a human impulse to promote symmetry and coherence in this universe.

Yet, our collective practical success in engineering, manufacture and to some extent in politics has encouraged a contrary idea that now threatens our existence. This is the perspective that 'nature' (and by this we have traditionally meant the non-human aspects of the Earth as we experience it) is an objective 'other'. Many of our most dominant political and business governors have found it useful to regard 'nature' not as our nurturing co-partner in evolution, providing in its inherent symmetry the impetus for our moral and social organisation; but as an entity valued chiefly for its seemingly inexhaustible resources. This attitude looms as a significant adverse factor in the contemporary regulatory context in which nanotechnology is becoming a global phenomenon.

For some such leaders a corollary of the view that our technological developments and ethical and religious understandings are outcomes of humanity's evolutionary struggle against 'nature' is that it's not only necessary, but also sufficient to examine human genes, hormones and neural synapses to discover the material source of every virtue, moral belief, or legal norm that has facilitated humanity's to-date successful contest for survival in this world.

The increasing specialisation of scientific research about 'nature' (including the unwillingness of scientists to seriously consider problems outside their metaphoric disciplinary silo) has meant that morals are increasingly regarded in that sphere as a matter largely of private concern and law as related chiefly to issues such as grant funding rules, patents, occupational health and safety requirements, politics, traffic rules, taxation or the divorce courts (not necessarily in that order of priority). Many nanotechnology scientists thus believe that a rigorous commitment to objective methods in order to discover truth provides adequate virtue to justify their lifetime of professional endeavour. Their duty, they might say, is to reveal nature's truths within the intellectual and collegial constraints of their discipline, and dedicated performance of that duty brings its own rewards in terms of good character.

Nonetheless, to reprise our emerging theme, science in so many fields underpinning nanotechnology is not only confirming the complex way in which 'nature' remains our nurturer, but is also revealing physical laws that prove 'nature' is structured around geometry and mathematics in ways diverging from the mental images of it we create by sensory information and common reasoning. Kurt Gödel's proof that there will always be some unsolvable mathematical proofs is but another example revealing that uncertainty is also an embedded 'law' of nature. The physicist John Wheeler has even argued that the laws of physics, like social laws, are evolving in step with our understanding of them. Is it only our limited temporal and spatial perspective that prevents us seeing how our social laws and our governance arrangements locally and globally are likely to emerge from our scientific attempts to discover, replicate and enhance patterns of universal symmetry?

There is a pressing need for such a fresh governance approach. At present the combined effect of many aspects of the advanced technology, as well as the social systems we've developed over the last few hundred years, threatens to destroy us as well as our biosphere. Unfortunately this is not an unwarranted, alarmist claim.

The critical global public health and environmental problems of our age (those we have bequeathed to so many future generations) include: increasing global population and demand for energy from old photosynthesis fuels (for example coal, oil and natural gas); a non-localised economy whose lack of social responsibility is exacerbated by corporate and governmental corruption as well as by democratic unaccountability; extreme and unpredictable weather events (floods, cyclones, droughts), rising sea levels and ocean acidification.

Disruptive climate change is being driven by higher atmospheric levels of human-produced, solar heat-trapping CO_2 (390 ppm) than the world has ever experienced (Arctic ice core studies going back 800,000 years show a range of 180–300 ppm). Other causes of anthropogenic climate change include increased emissions of methane (CH_4) and nitrous oxide (N_2O). Warming of high-latitude northern hemisphere areas and the Antarctic peninsula is changing the climatic temperature gradient, decreasing solar heat reflectivity and disrupting (perhaps beyond the point of recovery) the global carbon cycle, as land and oceanic 'sinks' become overloaded. There are also closely related challenges of massive biodiversity loss, degradation of ecosystems, as well as famine, poverty and inequalities in provision of the basic preconditions of life including security and access to basic food and essential medicines.

One proposed way morals and law may assist in remedying these issues involves setting 'planetary boundaries', or global governance principles designed to specify parameters in the natural world consistent with a safe operating space for humanity as assisted by its technologies (particularly in future by nanotechnology). These might include limits on human population (non-coercively, for example, through increasing educational opportunities for prospective parents), constraints under international and domestic law on atmospheric temperature and greenhouse gas levels, subsidies for renewable energy, as well as specifications of the amount of water and vegetation necessary to undertake photosynthesis across the globe.

At present, however, the idea of such 'planetary boundaries' is far from being incorporated in binding legal obligations (breach of which justifies damages or sanctions) in national statutes or international conventions. Indeed, for many people such 'planetary boundaries' seem now (in the context of existing global governance arrangements) to be as much idealised, paradoxical and anomalous entities as once were atoms, or are presently the notions of string theory and parallel universes.

This book explores the apparently anomalous idea that implementing such critically important environmental limits to human growth and exploitation of natural resources will necessitate a reconsideration of the basis by which global society should support laws (external constraints) mandating use of emerging technologies (such as nanotechnology) to promote sustainability of the natural environment. The mechanisms involved, as we shall see, not only may involve converting environmental 'boundary' principles into rights enforceable in national and international courts, but also fostering community, individual, corporate and governmental utilisation of public purpose-developed nanotechnology.

Such efforts to promote nanotechnology in the context of environmental sustainability could be viewed as a Quixotic modern day *psychomachia* – a naïve 'techno-fix' dressed up as a global contest of virtue against vice. Organisations such as Greenpeace, or the Ecovillage and Slow Food movements, as well as the Greens political parties, for example, though highly motivated to shape norms of sustainability in our governance arrangements, remain far from convinced that nanotechnology is or will ever be of great value in creating a sustainable world.

It should be clear now that a backdrop to our exploration of global governance strategies for nanotechnology involves examining whether there are necessary synergies between the collective enterprises of physical law-discovery and societal law-making. What we mean by a 'law' in this context will be one of the recurring questions examined in this text. For example, if we say that a concept is a 'law' do we mean we expect it can truthfully predict outcomes and has never yet been proved false, or that we must accept it is a law because others with authority (be they scientists or lawyers) have told us it is?

To summarise, an important aspect of our sustainable future is likely to reside not only in conceptually unifying all the laws of physics, but exploring the hypothesis that such a scheme is also coherent with moral theory and jurisprudence as well as widespread use of nanotechnology. Central here is the thought-experiment of considering physical and social laws from the standpoint of eternity, a perspective we'll investigate in greater detail in the next section.

1.2 GLOBAL NANOTECHNOLOGY GOVERNANCE FOR ETERNITY

I was fortunate enough to be born and grow up in one of the most naturally and artfully vegetated cities in the world – Canberra. Canberra, the national capital of Australia, was designed to nestle beneath the forested Brindabella Mountains by the vegetarian architect Walter Burley Griffin in the early 1900s. Griffin had a very difficult time convincing the bureaucrats involved in constructing the city to incorporate his futuristic ideas about urban dwelling in sustainable harmony with nature. Griffin's perspective of planning a city fit for eternity didn't gel with men accustomed to justifying their worth by the balancing of the next budget. The benefit to science and governance of adopting a longer view is the theme of this section.

My family continues to live in this beautiful city. Each day (particularly since we can view the growth of the National Arboretum from our front window) Canberra's landscape reminds me of the central tenet of ecosystem science – that sustainability in all of its formulations is critically dependent upon photosynthesis.

Photosynthesis has been building up food, fuel and oxygen on Earth for 2.5 GYr, since a time known in geological circles as the great oxidation event (GOE). Photosynthesis in one view may be regarded as the Earth breathing. It accounts for a global annual CO_2 flux of 124 PgC/yr and an annual O_2 flux of $\sim 10^{11}$ t/yr. Photosynthesis globally traps around 4,000 EJ/yr solar energy, in the form of biomass.

Photosynthesis generates carbohydrate from the carbon dioxide whose industrial production and atmospheric accumulation is dramatically altering our climate systems. It thus provides the primary food and energy source (in forms of 'old photosynthesis' – such as oil, coal, wood and natural gas) for human occupation of the world's ecosystems. It also makes the oxygen we breathe (by using solar energy to split water) and creates the atmosphere that protects us from damaging solar irradiation. The central importance of photosynthesis to human and environmental sustainability is an insight that global governance systems, particularly those related to new technologies (such as nanotechnology), have not yet properly appreciated or integrated.

Because of the centrality of photosynthesis to the sustainability of life on Earth (however we come to define that concept) it is worth briefly reviewing its key components. Photosynthetic organisms absorb energy from the sun in the form of photons (as Albert Einstein proved to win his Nobel Prize) but also waves in a particular band of the electromagnetic spectrum (~430–700 nm). They do this by utilising 'antenna' chlorophyll molecules in cell membrane thylakoids, or intracellular organelles called chloroplasts. The absorbed photons create an electric current that powers the oxygen-evolving complex (OEC) with the assistance of manganese (in the MN_4CaO_5 cluster) and a protein known as photosystem II (PSII) to oxidise water (H_2O) to hydrogen and oxygen (O_2) that is released to the atmosphere. In effect, bacteria and plants designed their own sustainable electricity supply billions of years ago – a point we haven't yet reached.

The electrons thereby produced are captured in chemical bonds by photosystem I (PSI) to reduce NADP (nicotinamide adenine dinucleotide phosphate) for storage in ATP (adenosine triphosphate) and NADPH (nature's form of hydrogen). In the 'dark reaction', ATP and NADPH as well as carbon dioxide (CO_2) are used in the Calvin-Benson cycle to make food in the form of carbohydrate via the energy-expensive enzyme RuBisCO (Ribulose-1,5-bisphosphate carboxylase oxygenase).

The photosynthetic system thus involves a tiny solar-powered electric current interacting in proteins to split water and combine the output with absorbed carbon dioxide. It has proven robustly capable of maintaining life on Earth for billions of years. Yet in its natural form photosynthesis is an outcome of the climatic conditions in which it evolved and is not that efficient. The average percentage conversion of incident light energy to chemical energy through photosynthesis, for example, is approximately 3 to 6 per cent with a theoretical maximum of about 13 per cent (much lower under suboptimal conditions such as low ambient light and reduced water availability). The interaction of nanotechnology with photosynthesis and governance concepts of sustainability is a base line running through this whole book that emerges as a melody in the concluding chapters.

Let's reprise the argument so far. We have provided introductory material on nanotechnology and the biologic process crucial to sustainability of life on Earth. We have demonstrated that nanotoxicology is an important regulatory issue. Indeed, some non-governmental organisations consider the unresolved problems of toxicity so great that global nanotechnology research and development should indefinitely be put on hold, leading to a deep future in which use of nanotechnology is carefully circumscribed, if allowed to exist at all. They argue, with considerable justification, that, although increasing attention is being paid to questions of safety by design, green chemistry and environmental footprint in nanochemistry laboratory work, the ecotoxicology of nanoparticles is still in its infancy.

Despite such concerns, we have shown that nanotechnology is already becoming a global industrial phenomenon. Its marketed applications are presently predominantly high profit-oriented consumer goods for the developed world. Yet (as will be examined particularly in Chapters 5, 6 and 7) with appropriate governance nanotechnology has significant potential applications in fields as diverse as energy supply, medicines, chemicals, manufacturing, food processing and military defence. We now begin to explore some of the key conceptual underpinnings that will permit nanotechnology through such applications to promote the social virtue of environmental sustainability as a foundational condition of global governance.

We have already argued that, to think properly about benefiting humanity now and in future generations, our policy-makers (both governmental and corporate) need collectively to adopt a perspective that has proven very fruitful intellectually in both philosophy and science. This involves viewing complex societal interactions and the anomalies they throw out as if from eternity, making logical extensions from proven laws even if that counters existing conceptions. Nanotechnology, for instance, when regarded from eternity, can be considered, like all our technologies, as a manifestation in time and space of our collective consciousness as shaped by material (matter-based) and moral (conscience-based) pressures upon it.

This book won't discuss whether or how to encourage humanity to adopt such a perspective (for example through a global culture promoting expansion of consciousness through altruistic service combined with contemplative detachment from memories that inhibit one-pointed concentration). Rather, it will examine what are the likely positive outcomes of global governance systems adopting such an extreme long-term approach to understanding and resolving complex policy issues. A major practical justification for the 'eternity' approach to global nanotechnology governance is that it is coherent with the emerging normative interest in the consequences of our actions for future generations and the sustainability of ecosystems.

In furthering our argument, we'll start in what may seem an unusual place. This is with Benedict de Spinoza's *Ethics*, specifically Bk II Prop. XLIV Coroll. II. Spinoza there writes that in terms of logic the best way to understand a thing truly is to try to perceive the idea of it under what he terms 'a certain species of eternity'. Spinoza might just as well have stated 'eternity or infinity', as his argument concerned the value of imagining such outer limits of time and space as our hypothetical conceptual vantage points in moral decision-making. Phrased in contemporary language, Spinoza claimed that making moral decisions from an eternal vantage point stimulates our powers of reason to detect patterns of symmetry and harmony that encourage us to transform personal self-regard into empathy and altruism.

Immanuel Kant propounded a similar view in his *Critique of Pure Reason* when he stated that logically space and time should not be considered as derived from temporal sensory experience, but as preconditions to our perception and thinking. In his *Introduction to the Groundwork of Metaphysics of Morals* Kant applied this insight about physical 'laws' to expound the view that the 'universal' viewpoint is the correct one from which to view the rightness of a moral decision. We should strive, Kant wrote, to apply moral principles that are capable of universal application because in doing so we achieve the only possible unqualified good in that universe.

That we view the world truly from eternity seems a rather abstract recommendation (fit for moral philosophy but not hard physics). Yet this is not true when we consider the many physical laws (many directly underpinning nanotechnology) discovered by philosophers and physicists such as Pythagoras, Kepler, Galileo, Newton, Maxwell, Einstein, Bohr, Schrödinger, Dirac, Witten, Calabi and Yau, by adopting just such an approach. To give but one example, both the discovery of astrophysical black holes and the development of string theory (as a means of unifying the equations of quantum physics and general relativity) are indebted to the geometric insight that, because the curvature of a sphere is inversely proportional to the radius squared, as the radius goes to zero the curvature goes to infinity.

Modern physics appears to be confirming that eternity and infinity are hard-wired into the immediate reality of 'nature' as we know it. If we consider, for example, how what is known as the Higgs Field gives mass to fundamental particles in different regions of the universe, then what might appear as eternity outside such a region may be infinity at each moment within it. Further, when physicists try to mesh the equations proven to predict the activities of the very small (quantum physics) with the very large (general relativity) they end up with infinite probabilities. This is mathematically incoherent unless our world has many more dimensions than we are used to or, indeed, inhabits an infinite number of universes. The 'big bang' or initial low entropy state that commenced our expanding universe appears to have also initiated time from a preceding condition that might as well be termed eternity. 'Eternitists' even claim that according to modern physics past, present and future may well be equally real, though that is certainly not our common experience.

Expressing this point another way, our usual approach to comprehending 'nature' (and that includes developing technology and the making of moral choices about its sustainable use) involves applying what we consider to be our common sense as it routinely operates in the three familiar dimensions of space and one of time. By considering anomalies in such experience from the perspective of eternity (or infinity), on the other hand, we facilitate our reasoning reaching a richer and potentially more accurate understanding of 'nature'. Geometry and mathematics provide elegant ways this can be done (for example, string theory physicists can coherently write the geometry of 10 or 24 dimensional space though that confounds our common sense).

At the nanoscale (a billionth of a metre) these issues manifest most pointedly. It is at the nanoscale that the laws of classical physics (more readily conformable to sensory experience) begin to interact with the much more uncertain features of quantum physics. Techniques such as neutron scattering and transmission electron microscopy (used to elucidate particles at the nanoscale) produce data involving indeterminacy, wave-particle duality and many other problematic features of the quantum world. Picotechnology, as a hypothetical future technological manipulation of subatomic matter (gluons, quarks, super strings, the quantum fluctuations that give rise to the universe's large amount of 'dark energy') will involve such anomalous features of reality as even more prominent factors.

Fundamental physics involves the search for laws governing the elementary building blocks of matter. When one reads physics texts stating that equations run just as well backwards as forwards in time, that everything in the universe is tending towards loss of complexity and dissipation of energy, a person interested in conscience might ask 'Why should any decision ultimately be valued more than another?' Why does being good ultimately matter? When we manifest virtues by acting on conscience what are we adding to the universe that wasn't there before? Is it a form of energy, or something else? If it is something else, what is its relationship with matter and energy, time and space?

Modern physics has shown (for example through the uncertainty principle) that an observing consciousness does alter fundamental states of matter and energy in space-time. Perhaps those fundamental states are designed to promote a special type of consciousness that we are striving towards in the guise of morality. The insight that reality (including human moral purpose therein) appears most clearly when viewed at the extremes of sensory experience (eternity and infinity) suggests a hypothesis that physics is working towards a unified theory not just of electromagnetism, the strong and weak nuclear forces (quantum mechanics) and gravity (general relativity), but also (as Kant supposed so many centuries ago) the moral law within us.

How does this apply to the use of new technology such as nanotechnology to create a sustainable world? Thinking properly about our responsibilities to future generations must involve considering how to preserve essential systems long term. Eternity is as long term as you can get. Industrial and technological revolutions resembling that involved with nanotechnology have occurred before, in more recent times characterised by a mindset that the natural resources they require (for example coal and oil) were infinite or could be utilised without much consideration that they were exhaustible.

The finite legacy of millions of years of photosynthesis provided the coal and oil that powered blast furnaces, steam and internal combustion engines, as well as making fertilisers for mass-production agriculture. The resultant material success promoted models of human prosperity based on ideological commitment to endless economic growth, technological innovation and global trade in manufactured goods and food drawing upon comparative advantages in abundance of raw materials, technological excellence and cheap labour. This created vast urban populations divorced from routine association with the natural world that they presumed would somehow indefinitely support their existence. The thinking of men who dominated these technological revolutions, in other words, tended to prioritise short-term profit and disregard long-term perspectives. The global governance structures they spawned reflected this focus – nation states and corporate globalisation in large measure drew their strength from increasing utilisation of raw materials taken (often with the assistance of military force) from the environment at the threat to sustainability of local communities.

Policy-makers continuing to embrace the corporate globalisation model of social development are likely to view as risky and threatening ideas that humans should try to use any new technology to sustain the environment as a dominant governance focus. The notion that the prime focus of our newest technology innovation (nanotechnology) should be to facilitate the social virtue of environmental sustainability probably seems to them unnecessarily idealistic and threatening to the social fabric. Yet it is important as a conceptual foundation of global governance structures to establish that this linkage of nanotechnology and sustainability is the outcome of a very logical process of reasoning.

The policy debate about environmental sustainability when viewed in the short term may seem a values contest about the use of the world's natural resources between proponents of public and private interests with no obvious chance of reaching substantial or imminent consensus. This, however, when viewed from a perspective we have conveniently described as 'eternal', is to misunderstand not only the mechanism by which humans make laws, but also the importance of jurisprudence making social laws coherent with those that underpin nature – the laws of physics. We return to the point, in other words, whereby idealist approaches to developing governance norms (such as human rights or environment protection legislation) may parallel methods in the physical sciences that utilise thought-experiments in which the observer's vantage point is eternity or infinity.

Jurisprudence is the field of knowledge traditionally concerned with defining societal 'law'. Mostly this law-defining appears to represent an alternation between what we may describe as the idealist perspective of eternity or infinity (responsible for inspiring many constitutional and international law human rights pronouncements) and, on the other hand, social organisation ideologies maximising for example the sovereign interests of Western nation states or the financial concerns of their major private interest groups (increasingly including supranational corporations).

As is the case in many other areas of human life, jurisprudence scholarship thus reflects the opposition between scholars, judges and politicians embracing the perspective of eternal ideals such as justice, equity and sustainability and political, judicial and corporate leaders more content with the conservative view that society should reflect virtues such as security, predictability and certainty (for example as embodied in the rule of law) that allow individuals the freedom to strive, acquire and prosper according to their own lights. Something inherent in human society (at least when it operates as a functioning democracy rather than as a military dictatorship or corporate resource) seems to drive this tension between progressive and conservative governance philosophies, making it as necessary as positive and negative poles for electric current or magnetic fields.

Physicists or biologists are tempted to denigrate jurisprudence for the imprecision and lack of objective verifiability of the 'laws' it considers (and economics with greater justification, but that may be a personal prejudice). Yet they should consider how speculative and vulnerable to ridicule some early formulations of physical 'laws' have been (Albert Einstein, for example, did his PhD to prove whether atoms existed and what size they were). Scientists should more frequently imagine in what peril they and their work are placed if the rule of law (the support of which arguably is a central task of jurisprudence) breaks down – as it has in many failed states today. In those disrupted nations governance by rules is replaced by arbitrary arrests, torture and executions, corruption and abuse of power, often cloaked in nationalistic propaganda, religious fundamentalism, or free market ideology.

Let's recall the key components of the argument so far. The perspective of eternity is a mathematically and geometrically robust route to truth in physics. The application of a similar technique in regard to the nature and purpose of putative social laws is generally characterised as an idealistic, rather than pro-symmetric process. When social laws about sustainability are considered this way it is likely to be erroneous to regard their central purpose solely in a self-interested or anthropocentric manner. The key word here is 'solely'. On this eternity approach, the appropriate way to view conservative political and legal positions may be as particles, while idealistic ones can be considered waves.

In recent years substantial numbers of idealistic humans have fought for the legal rights of other species and ecosystems. Likewise, scientific research increasingly confirms the critically threatened interdependence of life, and modes of rapid global communication like the Internet appear to be magnifying public awareness of this. Such examples provide indirect evidence of an increasing collective interest of humanity in championing norms about environmental sustainability in human governance relationships. Yet those opposing such 'wave-like' positions propose equally valid arguments supporting fiscal responsibility and societal stability.

Let's consider more closely how the concept of sustainability (emphasising future generations) is emerging in global governance systems. Environmental sustainability (chiefly focused on in this book) has received much less attention in this context than the more widely debated notion of developmental sustainability. The Brundtland Report in 1987 influentially defined 'sustainable development' as 'development that meets the needs of the present without compromising the ability of future generations to meet their own needs'. Sustainable development, however, is a policy concept that places the interests of humans at its core. The notable economist Robert Solow in his *An Almost Practical Step Toward Sustainability*, likewise, defined sustainability as a social virtue arising from consistent application of the ethical principle that the next human generation must be left with 'whatever it takes to achieve a standard of living at least as good as our own and to look after the next generation similarly'.

In recent years, legal documents such as the *New Delhi Declaration of Principles of International Law Relating to Sustainable Development* have highlighted the need to develop governance structures that facilitate sustainable use of natural resources and precaution against their damage or loss. The International Law Association (ILA) is actively investigating how to embed such principles in the broad framework of international law as applied, for example, by the International Court of Justice (ICJ), the International Tribunal for the Law of the Sea (ITLOS), human rights courts at national, regional and international levels, World Trade Organization (WTO) dispute settlement arbitral tribunals, inspection panels of multilateral development banks and compliance committees of multilateral environmental agreements.

Yet the jurisprudence of developmental sustainability is becoming closely linked with ecosystem science. The Intergovernmental Science-Policy Platform on Biodiversity and Ecosystem Services (IPBES), for example, plans to conduct periodic assessments of Earth's biodiversity and 'ecosystem services' – ecosystems outputs, such as fresh water, fish, game, timber and a stable climate, that benefit humankind. These assessments, based on reviews of the scientific literature, aim to answer questions about how much biodiversity is declining and what the implications of extinctions and ecosystem change might be for gaps in global governance and research on issues such as land management. The Nature Conservancy and the World Wildlife Fund have been working with academics to draw upon such research to develop the concept of 'natural capital' so that ecosystem 'services' (such as flood protection, crop pollination and carbon storage) can be valued in dollar terms and included in the economic assessments of business, community and government decisions.

Further, influential academic works such as William Vogt's *Road to Survival* (1949), Rachel Carson's *Silent Spring* (1962), Garret Hardin's *The Tragedy of the Commons* (1968), Paul Ehrlich's *Population Bomb* (1968) and Stephen Morse's *Sustainability: A Biological Perspective* (2010) have emphasised the critical challenge of ensuring that our world continues to involve harmonious interaction between all life forms. Principle Four of the *Earth Charter* similarly enunciates ethical obligations to (a) recognise that the freedom of action of each generation is qualified by the needs of future generations and (b) transmit to future generations values, traditions, and institutions that support the long-term flourishing of Earth's human and ecological communities.

The notable economist Amartya Sen in his *Idea of Justice* is another of those striving to expand the concept of sustainability beyond economic considerations so that it includes issues about access by future generations to political and legal freedoms and capabilities. Sen questions the idea that sustainability must be a human-centred virtue, one that has 'typically been defined in terms of the preservation and enhancement of the quality of human life'. Equally important have been theorists such as CS Holling who have advocated adaptive governance for sustainability in the context of promoting emergent complex system properties such as resilience and discontinuous changes in social response to new technology.

To summarise the argument in this section, scientific research confirming our interdependence with 'nature' is a major factor driving the collective conscience of humanity and then its governance systems to embrace the social virtue of environmental sustainability (as scientific theories such as evolution earlier did with governance virtues like justice, equality and liberty). This virtue when considered truly (as a species of 'eternity' as Spinoza put it, or from a 'universal' perspective as did Kant) and not contingently (for example as something useful to this particular generation or its dominant interest groups) may represent a planetary acknowledgement of fundamental interconnectedness that eventually will be explicable according to physical laws. Such a vision is behind projects such as the EarthSeeds project of the Spaceship Earth Foundation, which plans to place a picture of Earth from space in every classroom in the world.

Perhaps nanotechnology, complemented by breakthroughs such as the conceptual unification of the physical laws of quantum physics and general relativity, will provide a new point of reference for including social laws within such unification. This approach, emphasising individualised local (domestic and community) application of universal principles related to governance of nanotechnology for the global good of environmental sustainability, could even be referred to as 'nanogovernance'.

The concept of 'nanogovernance' brings together and extends ideas that have underpinned my publications over the past seven years. These include not only consideration of how to justly and equitably regulate the development of new health-related technologies, but also the relationship of laws of physical reality to conscience, the rule of law, the social contract, free market globalisation, state sovereignty and international civil society. How to begin extending the idea of nanogovernance to the global level is the subject of the next section.

1.3   WHY A GLOBAL NANOTECHNOLOGY FOR SUSTAINABILITY PROJECT?

As mentioned, when people talk of regulation of nanotechnology these days they are usually debating methods to investigate and control its potential toxicological impacts. My first exposure as an academic to the concept of regulating nanotechnology arose from a decision to help Richard White, a man who'd contracted silicosis from sandblasting fuel tanks at an Australian airforce base. Richard was dying from his disease, yet he wanted to leave a legacy that might assist other sufferers of similar toxic exposures. Together we helped set up a parliamentary enquiry into workers' compensation from injury by toxic dusts. At the last moment the terms of reference were altered to include nanoparticles, an area of research previously unknown to me. Sometime later (after many publications and conference presentations in the area) I received an Australian Research Council grant to study nanoregulation and became engaged in public policy debates over the potential toxicity of nanoparticles in sunscreens. This is a big issue in Australia where skin cancer is so prevalent. My research also explored ways in which our medicines' cost-effectiveness regulatory system could scientifically examine the community 'health innovation' of nanomedicines and how nanotechnology might impact on questions of national security.

In the course of this research I became surprised at the extent to which the pre-eminent regulatory challenges for nanotechnology were expressed in terms of time-framed goals for developing instruments to assess air and water exposure of organisms to engineered nanoparticles, as well as methods and strategic programs to evaluate their toxicological impact. Surely, I started to think, we should spend more time focusing on how this wonderful new technology can help us solve the potentially catastrophic public health and environmental challenges we now face.

This is not to say that nanotoxicology isn't a serious problem. Regulators and policy-makers have known for many years that many ENPs cause cellular toxicity, though the precise mechanism for this remains largely unknown. Human risk analyses and the principles and rules developed in conjunction with them are not easily extrapolated to strong *in vitro* evidence that ENPs elicit reactive oxygen species that cause cellular and DNA damage.

Regulators also now know that inhalation of particular types of ENPs such as multi-walled carbon nanotubes (like asbestos – relatively long and biopersistent fibres) can produce life-threatening problems (for instance a chronic inflammatory response resembling mesothelioma). Yet although occupational health and safety standards remain substantially underdeveloped for carbon nanotubes, research and business communities continue to invest heavily in these nanomaterials for a wide range of consumer products under the twin assumptions that they are profitable and no more hazardous than graphite.

If we wait for nanotoxicology research to conclusively prove that every aspect of ENP use is unambiguously safe, the nanotechnology revolution may stall and fail to reach what can be described as its moral potential. It is likely, for instance, that, even after further decades of nanotoxicology research, no regulatory agency will possess adequate data to completely back up their standards or methods to monitor ENP exposure risks, either in the laboratory, manufacturing processes, workplace or home. It is a reflection of this scientific uncertainty that, though some health technology and chemical regulators now specify distinct safety regulations/requirements that must be met by manufacturers of ENPs, most manufacturers or lab managers have no legal obligation to tell workers or researchers they are working with ENPs, to keep sufficiently detailed OH&S data sheets that are nanospecific, or to report ENP-related hazards.

Important regulatory systems for ENPs do need revision. Existing models focus on triggers for action related, for example, to amount manufactured per year, historical toxicity of a related macroscale substance and classification as a new chemical entity (not routinely used for nanotechnology). The risks of nanomaterials, however, are more likely to be determined by particle number, surface structure and surface activity including the fluxing corona of proteins that encrust nanoparticles when they enter biological systems. Such problems are compounded by the fact that nanotechnology is no longer developing in a context of local experimentation, but has emerged as a globally pervasive system that challenges both trial-and-error established toxicological protocols and existing models of national technology regulation. All this makes nanotoxicological research a viable candidate for coordinated global scientific and governance effort. Such a global project, however, will not of itself help solve the critical public health and environmental problems humanity faces.

Exceptions to the predominant focus on nanotoxicology in nanotechnology regulation have begun to emerge. *Nature Nanotechnology*, for instance, in 2007 contributed two articles to the Council of Science Editors' *Global Theme Issue on Poverty and Human Development* involving scientific journals in 37 countries. The articles focused on nanotechnology for clean water and the export of rare metals for nanotechnological applications. A related editorial began: 'It is easy to see nanotechnology as something that is being funded exclusively for the benefit of the developed world.' In urging a more altruistic vision, the editorial cited a survey of experts that claimed nanotechnology was capable of positively contributing to achievement of the United Nations *Millennium Development Goals*. The three leading applications were energy storage, production and conversion, agricultural productivity enhancement, and water treatment and remediation. What these editors were saying, in other words, was that nanotechnology needed to symbolically and practically link itself in the popular conscience with coordinated scientific efforts to help resolve some of the critical problems facing humanity and its environment.

This text not only supports such arguments, but seeks to develop them in the normative ('rule-making') realm for practical global applications. It makes the case that there are moral and human rights obligations upon the nanoscience and nanotechnology communities in particular to address global problems such as access to shelter, fuel, food and water. More unusually, it also proposes that the process of reasoning resulting in the discovery of the physical laws underpinning nanotechnology itself supports such ethical and legal obligations.

We should remind ourselves about some of the critical problems this world is facing. Access to shelter, fuel, food and water are central survival issues for those living in poverty and will be exacerbated as global population grows towards 10 billion by 2050. By that time world energy consumption (which varies greatly on estimates, for example, by the United National Development Program or the International Energy Agency) will have risen from 400 to over 500 EJ/yr. At the same time the predicted 2–3°C increase in atmospheric temperature and water vapour will have had unpredictable and likely adverse impacts on the yield of staple crops, even in a CO_2 rich atmosphere.

Because it is so central to the arguments that follow, it is now also necessary to take a little time reviewing some key scientific issues behind humanity's energy crisis. At present, daily power consumption for a citizen of a developed nation is about 125 kWh/day (~250 kWh/day for an adult person in the United States); much of this power being devoted to transport (~40 kWh/day), heating (~40 kWh/day) and domestic electrical appliances (~18 kWh/day), with the remainder lost in electricity conversion and distribution.

The kilowatt-hour (kWh) is a standard unit of electricity power production and consumption. One kilowatt-hour equals 3.6×10^6J and one 40W light bulb constantly switched on uses 1 kWh/day from the electricity grid. To understand what this means it is important to note some physical laws responsible for defining *power* as the rate at which *energy* is used. One joule of energy is equal to the work done when a force of 1 newton moves its point of application 1 metre. One newton represents the force which will give a mass of 1 kg an acceleration of 1 metre per second per second. One watt of power equals 1 joule/sec (1 kilowatt = 1000 watts). Recall that the average citizen in a developed nation uses 125 kWh/d. Exajoules per year (EJ/yr) (exa (E) = 10^{18}) is a measure of global power supply and consumption (the current estimate being over 400 EJ/yr).

This is crucial information relevant to our individual and collective energy 'footprint'. It highlights how central to human survival the problem of energy is. Our use of what may be described as 'old-photosynthesis' fuels to supply this energy is causing loss of a significant number of species not only on Earth, but also in the oceans as a result of acidification. It is possible that if Earth remains intact the numbers of some species may regenerate (possibly with the help of genetic engineering). That, however, will be a complex and difficult task unlikely to recreate all the marvellous complexity reflected in evolution of life on Earth over millions of years. If humanity survives without its present technology (a frequent occurrence in human history) the reserves of oil and coal we are now rapidly depleting may be critical to the recreation of civilisation.

It is reasonable then to assume that if nanotechnology is to assist resolving the great public health and environmental problems of our age it must be relevant to our energy crisis. The political will of nation states to look to new technology for energy solutions is manifest in international agreements such as the *Copenhagen Accord* (2009), which included a moral commitment to support renewable energy initiatives through an adaptation safety net (US$30 bn for 2010–2012, rising to US$100 bn per annum by 2020) and the *Copenhagen Green Climate Fund* supporting a technology transfer mechanism 'to accelerate technology development and transfer […] guided by a country-driven approach'.

The most obvious source of renewable power that nanotechnology may make more readily available is the gigantic nuclear reactor that our Earth orbits. The annual solar energy intercepted by the Earth at ~1.37 kW/m² is 5.5×10^6 EJ/yr (although the annual solar energy reflected by the atmosphere back to space at ~0.3 kW/m² is 1.6×10^6 EJ/yr). Thus, the solar energy potentially usable at ~1.0 kilowatts per square metre of the Earth is 3.9×10^6 EJ/yr. This is exponentially more than our current (400 EJ/yr) or projected (500 EJ/yr by 2050, 1000 EJ/yr by 2100) global levels of human energy consumption.

Raw sunshine at midday on a cloudless day, for example, can deliver ~1000 W/m² in mid-latitude regions and ~1200 W/m² in low-latitude dry desert areas. If we take into account the Earth's tilt, as well as diurnal and atmospheric influences on solar intensity, then this figure becomes approximately ~110 W/m². This produces a potential practically usable solar energy of between 1,500 and 50,000 EJ/yr. Photovoltaics (the capture of solar photons and their transmission to the electricity grid), as we'll examine in Chapter 8, is a major existing focus of nanotechnology research – displaying increasing efficiencies of energy capture and storage. Should photovoltaics then be the major global policy focus of nanotechnology and, if so, by what mechanism? What are the alternatives? How should global nanogovernance systems promote them?

The chapters that follow explore the reasons behind, the obstacles to, the likely candidates for and the governance structures of a global nanotechnology for sustainability (NES) project. Nanotechnology applications in clean water, food supply and medicines and renewable energy will emerge as prominent candidate topics for such a project, particularly because of the large amounts of existing government and corporate funding and research interest in these areas.

Some may detect evidence of convergent (or even emergent) global consciousness in the fact that scientists and policy-makers have an established record of coming together to plan and coordinate such macroscience projects focused on the achievement of global public goods. The Human Genome Project (HGP) is a notable instance of this type of 'Big Science' project that undoubtedly accelerated research in a crucial area for human health. Some such global science projects are focused in one place like the European Organization for Nuclear Research (CERN) or the international project on fusion energy (ITER). The HGP was not.

Appropriate governance arrangements are vital to their success. CERN, for example, involves an agreement between many nations to fund expensive new equipment (such as the Large Hadron Collider) open to use by independently funded physicists from around the world. ITER highlights the benefits of signatories undertaking in such an agreement to share scientific data, procurements, finance and staffing. As with CERN, the Hubble Space Telescope (funded by NASA in collaboration with the European Space Agency) allows any qualified scientist to submit a research proposal, successful applicants having a year after observation before their data is released to the entire scientific community.

The following pages advance more detailed arguments for a global NES project. Those who read stories to their children at night and like to see their own work as ultimately benefiting that next generation might like to view what follows as a treasure map.

The next two chapters thus may be imagined as depicting the conceptual travellers (for example personified as the characters Sustainability, Law, Conscience, Corporations, Humanity, Physics, Free Markets and the Environment) meeting in a tavern, hearing tales and deciding to go on a journey where much of the benefit arises from what they learn about each other and themselves along the way.

Chapter 2 allows us to meet the foundational governance concepts likely to underpin global NES. It explores, for example, the deep-rooted moral reasons for such a project. It considers what commonly agreed features should constitute a social law, or a sense of legal obligation requiring focus on nanotechnology in this way. In particular it focuses on how the less human-centred virtue of environmental sustainability should mesh with established social virtues such as justice, equality and respect for human dignity (that provide the influential theoretical foundations for many existing systems of legal obligation). It explores how such an endeavour critiques the intersections between basic legal ideals (such as the social contract and the rule of law) and the laws of physics underpinning nanoscience.

Chapter 3, to briefly continue the dramatic analogy, confronts us with those that may be considered the (often unwitting) villains of the piece. It introduces and explores some key global governance structures and actors likely to hinder nanotechnology from assisting to resolve our critical global problems. It examines, for example, many of the well-analysed reasons, apart from toxicological concerns, why policy-makers, regulatory scholars and corporate executives are reluctant to emphasise the role of nanotechnology in facilitating a sustainable world. Major barriers we'll scrutinise here include the interest of governmental oligarchies in protecting what they perceive to be their sovereign interests and the way international trade and investment agreements have created unprecedented leverage for supranational corporate lobbyists to oppose public good focused domestic health and technology policy.

Chapter 4 considers how widespread use of nanotechnology by future generations (in what is termed a 'nanotechnology-embedded world') may support the foundational social virtue of environmental sustainability. This section concludes by proposing certain criteria that might be useful in deciding what would be the best macroscience global NES project.

Chapters 5 to 8 apply these criteria to the strongest potential candidates for a global nanotechnology project most suited to promoting social virtues such as environmental sustainability and human flourishing. Chapter 5 looks at whether it would be best to work on a global project involving nanotechnology's contribution to food supply as well as safe drinking water. Chapter 6 considers whether nanomedicine, particularly the use of nanotechnology to improve pharmaceuticals and medical devices, might be a good candidate. Chapter 7 likewise considers the possibility for nanotechnology in connection with global peace and security. Chapter 8 reviews the prospects for an international nanotechnology project dealing with issues of climate change and renewable energy.

What these case studies are seeking to establish is where policy-makers should focus in seeking to discover a globally accessible, cheap use of nanotechnology that can rapidly be rolled out on a scale large enough to resolve the critical contemporary problems for human society and its environment. As the physicist Freeman Dyson pointed out, all we need to save the world is one cheap and successful technology, achieved before our capacity to do so runs out.

Chapter 9 sets out the case justifying a Global Artificial Photosynthesis (GAP) project as the most likely candidate to satisfy the proposed criteria for a global macroscience NES project. That chapter analyses the governance issues that will need to be solved to bring such an endeavour in planetary nanomedicine to fruition and critiques different models that might be employed. It argues that an NES GAP project can justifiably be regarded as the moral culmination of nanotechnology.

2. Nanoscience for a sustainable world: a goal or set of principles?

> [T]he most dramatic feature
> of the universe to have survived inflation:
> the three-plus-one dimensionality of space time ...
> exhibits a high degree of symmetry ...
> it is necessary to calculate the wave function of the universe
> just as we ordinarily calculate the wave function of an electron.
> – Heinz Pagels, *Perfect Symmetry*

> And when we describe things as 'taking place',
> We're talking like builders, who say that blocks in a wall or a pyramid
> 'take their place' in the structure, and fit together in a harmonious pattern ...
> Look straight ahead, where nature is leading you –
> Nature in general through the things that happen to you;
> And your own nature, through your own actions.
> – Marcus Aurelius, *Meditations*

2.1 ETHICS OF NANOTECHNOLOGY-BASED ENVIRONMENTAL SUSTAINABILITY (NES)

In 1981 I visited the USA for the first time as a participant in a law students' international law mooting competition called the Jessup Cup. The competition was designed to encourage young legal scholars to learn more about how international law can assist to resolve complex geopolitical issues – in this case a sea bed demarcation dispute between two fictional coastal nations. It involved deliberation of provisions in a United Nations convention specifying that the deep sea bed was the common heritage of humanity. 'Equity' was one principle to be used in the demarcation.

After the competition concluded there was one town I particularly wanted to visit. This was Concord, Massachusetts, where once lived Henry David Thoreau, the author of *Walden* and *Civil Disobedience*. I recall walking out one afternoon to Walden Pond and seeing the cairn of stones that marked where Thoreau's hut had once stood. One then recalled inevitably these words of Thoreau's: 'I went to the woods because I wished to live deliberately, to front only the essential facts of life, and see if I could not learn what it had to teach, and not, when I came to die, discover that I had not lived.' I began to think then about how important it is to listen to nature – not just for relaxation, but for career choices; that the future might involve what I was *meant* to do in equal measure with what I *wanted* to do.

One thing that has struck me since is the surprising number of researchers and policy-makers who are willing to go into new fields because of a sense that it is somehow what nature demands of them. I've come to view myself as in almost equal measure a physicist and a non-denominational Christian in terms of religion. Physics and the received doctrines of any faith seem to be part of the same human impulse to use what intellectual processes and technology we have available to understand our place in the universe.

As an academic I've developed a philosophy about trying to 'surf' waves of 'natural need' in my publications and grant applications. I've discovered this also involves being aware that just prior to any related potential career success you've prayed for there will arise (as if by exquisitely constructed coincidence) some moral temptation to be resisted. I've repeatedly tested this approach (of seeking to attune conscience to harmonics of nature) and now just accept it as an inspiring and enlarging part of the natural order.

The proposal for a global NES project is a direct outcome of such thinking. Such a project, if it is to be established, will involve scholars in ethical and legal as well as scientific disciplines investigating not only how their research can help sustain humanity, but all life on Earth. This is a departure from prior anthropocentric approaches to governance theory. If it is part of a 'wave' of natural need then others will perceive that. It might be that nanogovernance is how we 'tune' our societies to symmetry in the cosmos.

So what is 'environmental sustainability' and why should it be considered a primary social virtue, a pre-eminent moral or ethical consideration for a global NES project, particularly when it doesn't seem to prioritise the interests of the human species? Wouldn't 'developmental sustainability' be a more practical term? Related issues include: 'Why assert the necessary connection between environment sustainability and the development and transfer of our most promising new technology – nanotechnology?' Wouldn't behaviour change be a simpler option? Why hasn't more importance previously been given in governance systems to supporting the process of photosynthesis as the physical basis of environmental sustainability? This chapter aims to provide a conceptual grounding to answer such questions.

First we need to clarify what we mean by 'governance systems' in this context. Governance involves coordinating social activity through principles and rules – of ethics and law. One important task of scholars in ethics, for example, is to assist the development of considered thinking by the general population and policy-makers about how to value things and determine right from wrong. The core of this reflective activity involves formulating, or as is more often the case clarifying, universally applicable principles rationally capable of being applied consistently in the face of obstacles. National and international governance systems emerged from a process whereby such moral or ethical insights are gradually transmuted into published, officially recognised and enforceable laws.

Unpacking ethical debates can be a challenging governance task when related to complex and rapidly advancing areas of new scientific advancement such as manipulation of the human genome, synthetic biology and nanotechnology. Ethics experts called upon to explain some such new advance to the media often repeat stock phrases about 'playing God', or the need for a new committee as well as extra grant funding to research the issue. Ethics analysis is beginning to play a less muted role in media and policy debates about protecting features of the natural world vital to its sustainability (such as photosynthesis) that have hitherto largely been taken for granted as presumptively inexhaustible resources for human economic development. Part of the reason for the prior muting may be that some influential ethicists (such as the Australian Peter Singer) are reluctant to ascribe ethical status to life forms that we cannot comprehend as 'suffering'. In this section we'll examine how ethicists are catalysing the incorporation of environmental sustainability into traditional human interest-focused governance structures.

The terms 'moral' and 'ethical' are often used interchangeably, though the latter has stronger professional and institutional governance connotations. Ethical principles likely to underpin a global NES project would traditionally have been derived either directly from religious authority or indirectly through confronting the necessities of survival – an understanding that without such norms or rules of behaviour the majority of humans could not live well with each other, or for very long. It will be a difficult but important task to reconcile the religious components of ethical systems required to support or contribute to a governance framework for a global NES project.

One way this can be done involves investigating the hypothesis that the laws of fundamental physics complement the conceptual underpinning of ethical principles derived from the major religious traditions. Historical research and improved understanding between the faiths, for example, can be viewed as endorsing the proposition that the founders of religions may be regarded as the equivalent of scientists in their time – seeking and expounding what they considered to be the truth about nature and the human condition. The earliest of the world's three great religious figures, the Buddha, was most explicit in framing his primary task in such ecumenical, undogmatic and even 'scientific' terms.

This book endorses the proposition that, backed by increasing scientific evidence of ecosystem fragility and importance to our own survival, environmental sustainability is beginning to resonate in our collective conscience. The evidence for this includes the changing content of moral debates in the mass media and Internet-supported civil society. Even in academia, more bioethicists for instance are transferring their research focus from intractable private questions concerning the beginning and end of life, to issues not only of global health, but also of environmental sustainability.

But it is not easy to discern where sustainability (let alone environmental sustainability) sits in that loose confederation of published ideas constituting the academic discipline of ethics. One finds little if any reference to sustainability in ethical works derived from religious traditions, or from academic schools such as utilitarianism ('act on principles maximising the greatest good for the greatest number') or deontologic idealism ('act on principles capable of universal application'). It could, of course, be argued that the concept of sustainability is present implicitly in such doctrines or in core religious concepts like Buddhist compassion, Christian conscience or Islamic *taqwa*.

Every reputable system of religious ethics, for instance, supports the proposition that a moral, good or virtuous life cannot be led by a person who thinks only of their own interests, or is disrespectful of God's creation. Sustainability likewise in most formulations necessarily requires consideration of the greatest good of the greatest possible number of stakeholders now and into the future. Arguably, thinking of as broad a range of 'others' as possible that may be impacted by our actions has indirect personal utilitarian benefits through fostering continuing peaceful relationships. This understanding is so central to social success that even sociopaths (and by their legal construction that includes artificial persons such as private corporations) pretend to appreciate it.

It is difficult nonetheless to discern sustainability as a clearly defined ethical concept in contemporary media and policy debates between those with Christian, Islamic or secular perspectives; those enmeshed in securing the embellishments of institutional power and those critiquing the desuetude of that power in the face of moral crises.

To give one example, 'virtue ethics' is one area of ethical theory to hitherto pay little attention to environmental sustainability. Ethicists supporting what is termed a 'virtue ethics' theory claim that the central fact of ethics is not a group of principles and rules, but the character of the 'agent' involved. It could be argued that the 'virtue ethics' position supports the social virtue of sustainability because its achievement is one of the central altruistic goals of good people in our age. Such goal-oriented virtue-based approaches to ethics have been criticised as too readily subjugating individual liberties and flourishing to the attainment of ostensibly wider social aims. On the other hand, non-goal-oriented virtue ethics theories are commonly subjected to objections emphasising their circularity and failure to provide determinate guides to action. Alasdair MacIntyre in his influential work *After Virtue* also pointed out that attempts to solve ethical arguments through the development of virtue have become a problematic task for many contemporary societies because most now lack a commonly agreed end point for moral action.

Ethics may approach the less human-centred social virtue of environmental sustainability by extension from moral concerns in related areas. Philosophers such as the Australian utilitarian Peter Singer, for example, have argued that it is now time to restructure ethical thinking regarding animals and remove 'speciesism'. Singer advocates that the arguments against more human-based forms of discrimination, such as sexism and racism, are equally applicable to animals, based on their common capacity to suffer and the primacy of preventing suffering as far as practicable for the majority of our interactions.

Reasoning by analogy from how women and non-white races were previously understood by many white males to be inferior beings, Singer proposes that animals should also be given similar rights and protections as are now enjoyed by those minorities. The core of Singer's stance is that, once we recognise the capacity of life forms to suffer, ethical interests are invoked that should be protected. This focus on the normative primacy of 'suffering' creates conceptual difficulties for Singer when he attempts (as he does) to argue that ecosystems (such as wild rivers or rainforests) deserve ethical protection. Why should expanding the circle of empathy require proof that an entity can suffer because it possesses a nervous system comparable to our own? Utilitarian concerns about maximally relieving suffering may not be the most appropriate basis upon which the ethical foundations of a global NES project could be based.

Another approach to conceptualising the ethical foundations of a global NES project might involve extrapolation from ethical systems applying in areas such as health care and doctor–patient relations. If our planet is regarded as a patient (as can be argued for example by extension from the Gaia hypothesis of James Lovelock) then ethical norms concerning environmental sustainability can be developed by analogy from the system of basic ethical principles applying to more traditional doctor–patient relationships.

For example, Tom Beauchamp and James Childress in their influential work *Principles of Biomedical Ethics* claimed to have derived the four basic principles of medical ethics (autonomy, non-maleficence, beneficence and justice), not from the posited foundational virtues of any ideal or utopian doctor–patient relationship, but from 'considered judgments in the common morality and medical tradition'. This appeared designed to link their ethical 'principlism' to the supposed authority of descriptive sociological research, rather than natural law theory (which as we shall see might be a better fit for environmental sustainability).

In the system of Beauchamp and Childress, the bioethical principle known as *autonomy* was defined as 'respect for the deliberated self-rule of patients (or research participants)', and linked to Kant's 'categorical imperative' to treat humans as ends complete in themselves, with intrinsic dignity; not as means to other goods. Alternatively, the authors pointed out, *autonomy* could be defined from a utilitarian perspective as requiring a constraint on the principle of *paternalism*. As the English philosopher John Stuart Mill wrote: the only purpose for which power can be rightfully exercised over any member of a civilised community, against his [or her] will, is to prevent harm to others. His [or her] own good, either physical or moral, is not a sufficient warrant.

Extending this norm of *autonomy* (or respect for intrinsic dignity) to the planet as a whole may support the ethical principle that our world should be treated as a type of collected consciousness and an end in itself, not instrumentalised as a means to some other good (for example economic growth). Respect for *autonomy* appears to require some capacity to consult and follow the independent will of another entity. Could scientific ecosystem research or consultations with environmental organisations provide a means of facilitating a type of *planetary autonomy* or deliberative self-rule?

From the principle of *autonomy*, Beauchamp and Childress claimed that more specific ethical rules could be deduced. These included a professional obligation to keep promises made to patients (or research participants), to maintain confidentiality of their information, to tell them the truth about their treatment and to ensure competence and skill in communicating information and performing relevant procedures. Extending these rules to the planet as a whole would require a considerable expansion of our collective capacity to identify with ecosystems. Yet even this becomes practicable if environmental advocacy groups including non-governmental organisations are regarded as expressing the interests of the environment.

The principles of *beneficence* and *non-maleficence* were joined in traditional medical ethics in a duty to provide net medical benefit to patients with minimal harm. *Beneficence* was importantly additionally associated with the ethical duty to undertake research and participate in professional education and training. Beneficence likewise was held to be demonstrated through sensitivity to risk of harm, potential of benefit, welfare and interests of involved parties, as well as the ability to reflect on the social and welfare implications of research. The ethical principles of beneficence and non-maleficence can readily be taken across to the humanity–planet relationship. Indeed, global public health physicians could regard themselves as acting in accordance with such ethical obligations towards the sustainability of planetary ecosystems.

The principle of *justice*, in relation to health and medical research ethics, divides into three ethical obligations: to ensure fair distribution of scarce resources (distributive justice), to respect patients' rights (rights-based justice) and to respect morally acceptable laws (legal justice). Justice is deemed to be present in medical research where the benefits and burdens are fairly distributed and the recruitment of participants and review is procedurally fair. Justice is a foundational social virtue as well as a basic principle of medical ethics and also readily translatable to a focus on the interests of the planet as a whole.

The ethical argument, for example, can be made that anthropocentric climate change is fundamentally *unjust* in relation to future generations not only because of its potential devastating effects on their health and well-being (particularly of vulnerable populations amongst them), but also because of their limited opportunities to remedy it and the capacity of developed nations (that benefited economically from the industry causing the emissions) to remedy it now with relatively little additional suffering.

Significant arguments raised against phrasing the climate change issue in terms of *distributive justice* include the lack of reciprocity of future generations to us, the problematic ethical status of the 'interests' of future generations, the unpredictable numbers of people likely to be impacted and the potential attenuation of moral responsibility with increased remoteness in time. Most conceptions of what is *unjust* deal with adverse consequences that are not irrevocably vague through being remote in time and space.

One of the undoubted advantages of the 'four principles' approach has been the ease with which its components can be recalled and act as a simple mental trigger for complex duties. If these principles could be viewed as applying to the planet as patient, then this would be a strength in terms of public understanding of the ethical foundations of a global NES project.

The four principles ethics approach, however, has been significantly challenged in recent times in medical academia, potentially undermining its value to an ethics of planetary sustainability. Many medical educators now view demonstrated capacity to memorise these four ethical principles and recite them when asked like a caged cockatoo, as providing insufficient proof that a health professional could adequately reassure patients or research participants of being not only physically at their side, but in conscience 'on their side'. This highlights the importance of ethical foundations of a global NES project not becoming merely tokenistic 'window-dressing', but being formulated in such a way that the bulk of the human population is inspired to implement them in the face of obstacles.

It is important to emphasise, at this point in the argument, how important is the task we're engaged in here, of finding an appropriate ethical foundation for a global NES project. If such a project, as I've suggested, is part of some 'wave of need' then its ethical foundations are already being developed. The task, to adopt a chemical analogy, is to make that reaction occur faster in time to effectively respond to the critical public health and environmental challenges of our age.

If left to be haphazardly founded on increasingly vestigial religious-based moral theories, moral claims that nanotechnology should (or should not) ethically be focused on environmental sustainability might simply become another strategy used by different theologic interest groups to persuade each other that marginalisation from the key positions of societal governance is not their permanent or necessary condition.

The ethics of a global NES project might also be presented as an element of enlightenment theory, the notion that society is progressively evolving to express greater reason in its conscious organisation. The enlightenment idea that there is an evolutionary progression in human reason and conscience, intersects well with the notion that promotion of greater organisation and complexity seems to be a natural extension of life and consciousness in this universe. One might even suggest that conscience allows us to perceive in ethical rearrangements a geometric connectedness, an appreciation of the same type of mathematical symmetry and harmony science has established as arising from nature's physical laws. Indeed, a global NES project may find its firmest ethical foundations in such an ethical framework. I shall now outline the basis of this approach in more detail.

As mentioned earlier, in the seventeenth century the philosopher Benedict de Spinoza wrote in his *Ethics* (Bk II, Prop. XLIV) that it is the nature of reason properly applied to perceive things truly, that is, as they are in themselves not as contingently existing in past, present or future circumstances revealed to us by sensory experience. This pronouncement was often dismissed as a type of idealist rhetoric. How, for example, could it make us view things any closer to reality to regard them as not bound by forward-flowing time? Such a position was contradicted by our sensory experience. Yet Spinoza's profound realisation (and others like it) paved the way for major scientific as well as ethical breakthroughs in thinking.

In the eighteenth century, Immanuel Kant (the philosopher central to enlightenment theory) similarly influentially argued that the capacity to form ethical concepts in the form of goals or end points for future actions based on universally applicable principles is a core distinguishing characteristic of the well-developed human mind. It arises, he maintained, proportionally with our capacity to view the world more objectively, including viewing ultimately our understanding of time and space as arising a priori as necessary preconditions for sensory experience (rather than being determined by it). Kant made these claims particularly in his *Groundwork of the Metaphysics of Morals* and his *Critique of Pure Reason* (though philosophers tend not to grasp the linkage between these positions). The freedom of individuals to set conceptual goals presupposes a capacity to reject them and Kant reasoned that laws (backed up by official enforcement) provide an external constraint upon persons whose selected end points would otherwise unduly interfere with the capacity of other rational beings to choose their own goals.

Kant's was an optimistic moral philosophy about human nature. Ethics Kant saw as rationalising voluntary self-constraint by minds that were seeking to embrace universally applicable principles. The more people acted from a concept of duty (often against the opposition of their own sensual inclinations) to consistently apply such principles, the more humanity was morally developing towards a type of collective enlightenment. Those principles also rose in ethical value the more they facilitated the capacity to flourish equally in all other rational beings. Kant summarised this by stating (in his *Introduction to the Doctrine of Virtue*) that virtue arises from consistent voluntary decisions to act (despite internal or external obstacles) upon principles capable and worthy of application by all rational humans. The use of new technology such as nanotechnology to support sustainability of human development fits squarely within this influential moral framework. Sustainability of the environment (as a non-rational thinking entity) at first glance has a more uncertain moral place within it.

Yet Kant, like Spinoza, paved the way for both the science of physics and governance systems involving ethical principles and legal rules, to progress in new directions. In his *Critique of Pure Reason* Kant wrote that 'we can never represent to ourselves the absence of space, though we can quite well think it as empty of objects'. Likewise 'appearances may one and all, vanish; but time (as the universal condition of their possibility) cannot itself be removed'. The twentieth-century physicist Albert Einstein, who was exposed to Kant's ideas in his youth, probably drew upon this insight (that space and time might exist in ways that seem at odds with everyday sensory experience) to ponder why the speed of light is constant regardless of the speed of its source and create his special and general theories of relativity as an answer to this apparent anomaly.

A corollary of such 'pure' reasoning, as Kant perceived, was that knowledge (including moral truths about the role of principles and virtues in constraining free will) could also arise from a suprasensible part of nature that has the potential to be true, despite not correlating with common experience. Such realisation may have been a critical factor in development (particularly by earlier enlightenment philosophers such as the physician John Locke) of the concept of inalienable human rights (granted by nature to all people) even though such a position had no foundation in the wider sociological facts of the time.

It is difficult now to grasp how important an advance it was in terms of the sustainability of human civilisation to claim that the basis of ethical and legal obligation could reside in an ideal applicable to all people as part of the 'nature of things' rather than be entirely constrained by the interests of a king or religion. Perhaps just as major an advance will be incorporating environmental sustainability into our constitutional arrangements.

To summarise, we have explored conceptual frameworks by which moral or ethical principles may underpin a global NES project. It is important to note that, just as with individual humans, societies develop and maintain virtues through commitment to consistently implement such principles in the face of obstacles. The great social virtues traditionally include: justice; equality; respect for human dignity; amongst the scientific professions, respect for truth; and in the healing professions, loyalty to the relief of suffering. As we'll investigate in the next section, these ideals and their associated principles gradually metamorphose (after political debate and struggle) into specific legal rules and enforceable legal rights, the consistent and predictable application of which in turn sustains those foundational virtues.

This is an important way of considering the fundamentals of the global governance framework into which the virtue of nanotechnology-based environmental sustainability must fit alongside justice, equality and respect for human dignity. The conceptual linkage of nanotechnology with such a non-anthropocentric virtue will involve a profound reorientation for most ethical systems.

Such a formulation of nanotechnology-based environmental sustainability, of course, can also be viewed (and perhaps dismissed by those more interested in social virtues such as economic growth, predictability and consistency) as another arcane, overly idealistic, impractical and insubstantial formulation of how individuals and societies may lead the philosophical 'good life'. Some contemporary ethicists even hold that society is shifting from endorsing an ethics based on such systematically developed enlightenment ideals and universally applicable principles to an ethics based on local practices, convenience and suspicion of theory – a pseudo-ethics of 'getting away with it'.

Martha Nussbaum rightly claims that this 'attack' on virtue ethics represents a 'confused story'. It is confused, Nussbaum maintains, because it is an error not only to claim that the works of the central 'enlightenment' theorist Immanuel Kant reveal an obsession with idealised duty and principle to the exclusion of character-formation and the training of the passions, but also to assert that the scholarship of seminal utilitarians such as Henry Sidgwick, Jeremy Bentham and JS Mill took little account of the importance of ethical systems in developing individual and social virtues. One could add that it is also 'confused' because while our political and corporate leaders do appear to manifest an ethics of media 'spin' ('if you don't like those ethical principles – our focus group can find others') that may be less the case with Internet-based civil society.

Environmental sustainability as a primary social virtue is linked with what can be called 'ecocentric' or 'biocentric' ethics. This is also known by terms such as *Gaia Theory*, or *Deep Ecology* and expressed in documents like the *Earth Charter* or *Earth Manifesto*. It involves a set of moral or ethical principles that may be summarised as emphasising two key principles. The first is that the flourishing and diversity of non-human life forms has intrinsic value requiring protection by policies and technologies reducing the number of humans along with their demands on those other species. The second holds that human flourishing itself requires a deepening respect for right relations with ecosystems that should be reflected in the choices our species makes about the use of new technologies. Such philosophies, despite the caricature versions of them sometimes presented by their conservative opponents, do not advocate any now impossible (if only because of population size) return to pre-industrial or new technology-free lifestyles in what may be termed a type of global Amish-ism.

Finally, we'll explore the potential ethical foundations of a global NES project through the lens of economics, even though ethics and economics are often regarded as disparate areas of academic discourse (and practical reality, given examples such as the global financial crisis).

Attempts have been made by some economists to frame the ethics of sustainability in terms of the fictional notion of perpetual growth in gross domestic product (GDP). Such formulations often pay obeisance to the fictional power of deregulated markets and the 'invisible hand' of entrepreneurial self-interest to ethically regulate demand upon Earth's resources in a sustainable manner.

Yet other economists have striven to factor our moral responsibilities concerning the finite and fragile resources of the biosphere much more centrally into their economic calculations. The virtues of ecological sustainability and environmental integrity, for instance, were influentially propounded by eco-economists such as EF Schumacher (with his concept of 'small (and local) is beautiful') and Kenneth Boulding (with his idea of 'Spaceship Earth' as a closed economy requiring recycling of resources). In doing this, the former drew upon Buddhist principles and virtues, while the latter relied upon those resonating with the Quakers.

Other economists such as Herman Daly have drawn on the laws of thermodynamics and the central importance of entropy to champion the idea of 'steady-state' economics that financially values maintenance of ecosystems equally with production and profits. One could go further with such an approach and suggest that ethics and economic principles should be coherent not just with thermodynamics, but also with physical laws and patterns of symmetry like those underpinning electromagnetism, gravity, general relativity and quantum physics, as well as other principles that are non-falsifiable, without necessarily correlating with our sensory-oriented experience of the world.

Many economists interested in developing greater moral and scientific credibility for their discipline are investigating the ethics of sustainability through the lens of human population and ecosystem science. One such approach, for example, defines sustainability as involving the persistence of diversity and ethical ideas of human flourishing among human communities, as well as the preservation and regeneration of ecological systems. One should point out once again that it is quite remarkable how little attention has been paid in such discussions to humanity's collective ethical obligation to sustain the process of photosynthesis.

To conclude, however this is characterised in terms of academic ethics theory, the moral reasons for supporting a global NES project may resolve into the simple understanding that it offers a reasonable prospect of significantly addressing, with but a relatively small opportunity cost, problems for which this generation bears some culpability that are likely to cause suffering and death on a large scale.

The above discussion has highlighted some of the ethical considerations involved in proposing nanotechnology-based environmental sustainability as a primary social virtue that may underpin a global NES project. In the next section we'll explore how the notion of nanotechnology-based environmental sustainability may mesh with two foundational legal concepts of global governance: the social contract and rule of law.

2.2 NES IN A NEW GLOBAL SOCIAL CONTRACT

After finishing my term as associate to Justice Lionel Murphy at the High Court of Australia I practised law in two of Australia's largest legal firms, Mallesons and Freehills. In that role I made pleas in mitigation of sentence for young boys (by now repentant) who had vandalised their school science lab on a full moon night. I represented men and women prosecuted for driving under the influence of alcohol, wrote advice on insurance claims, assisted tenants to obtain redress against landlords, and drafted building contracts and large commercial leases. I discovered how practising lawyers help human beings resolve all manner of disputes using a pre-existing set of rules that can be read and appreciated by each side. In all this, one could discern a society functioning according to a social contract under the rule of law. In so many countries today corruption and abuse of power ensures this is not what happens.

The governance arrangements supporting a global NES project to be effective and widely supported must be conceptually underpinned by the well-established jurisprudential notion of a hypothetical social contract. The concept of a social contract in broad terms holds that a society's normative (or law-making) foundations are principles and rights emerging out of community respect for great social virtues related to how humans *should* treat each other in ideal conditions. Such foundational virtues are well recognised as including formal and distributive justice, fairness and respect for human dignity. The related principles and rights typically include topics such as freedom of conscience, expression and association free from state interference except as necessary to protect the same societal claims in others.

This hypothetical contract provides a symbolic point of reference against which laws and political actions can be critiqued and ultimately challenged by individual members of society. The governance arrangements of a global social contract supporting a macroscience NES project could be based on approaches such as multilateralism (groupings of nation states with the international organisations they fund), market governance (involving imperfect business competition by supranational corporations), grassroots globalism (global authority decentralised to self-governing local communities), multiple regionalisms (democratic decision-making units not organised as states), world statism (with planetary military security, a planetary assembly, a whole world executive, planetary court and planetary bank), networked governance (nation states providing basic functions supplemented by collectives of non-governmental actors) and institutional hierarchies (a multiplicity of individual, community and international actors).

Proposing nanotechnology-based environmental sustainability as a primary social virtue in such a global social contract is somewhat unusual because environmental sustainability and nanotechnology have received little attention to date in social contract theories, certainly not at the global level and definitely not as linked concepts.

The social contract is a governance theory for reconciling individual rights and freedoms with societal cohesion. It is a fundamental conceptual entity that can be viewed, for example, on the approach taken here as explaining constitutional arrangements balancing state power and personal liberty in a manner analogous to how string theory balances quantum mechanics and general relativity, DNA provides a template for the folding of proteins, or the Higgs Field gives mass to energy.

The social contract is predicated on the idea that a society whose citizens strive to live by principle rather than self-interest must impose restraints on the exercise of free will. For individuals with mature reasoning, conscience sets such limits voluntarily. When such restraints have to be imposed involuntarily, they can most readily be accepted as deriving from an ideal state of affairs when perceived as arising by consent of the whole community. Living by principles designed to treat the interests of all equally means determining actions by criteria that do not prioritise maximising contingent individual interests. It also thereby necessitates living together to create what may be described from a 'deep' perspective as a type of geometric symmetry in citizens' relations.

Often such a contract manifests as a written constitution administered by senior judicial officers. Constitutional rights (claims and privileges enforceable by citizens against their government or each other under the rule of law) represent an abstract attribution to citizens' interests derived from the existence of a social contract. Purportedly derived from perception of an ideal form by key law-makers, a citizen gradually becomes aware of such rights within his or her personality the way he or she does virtues, for example through our conscience when we 'sense' a contraction to a 'point' when our rights or those of our fellows have been violated and an expansion to a 'wave' as we support these claims by others. When such a realisation of 'contraction' becomes collective, persistent and overwhelming, a government may be rejected and the constitution entrenching it amended or even replaced.

Many think citizens learn of their 'rights' by understanding their society's constitution and written laws. These texts may be viewed as formal 'positivist' representations of a social contract, that is, as an outcome of protracted academic and political debates about governance structures. So-called constructivist theorists also point to the role of civil society organisation in urging the emergence of new legal rights and responsibilities in a process termed normative diffusion.

From the 'deeper' perspective advocated here, on the other hand, such organisational structures are forces responding to enhanced awareness of more fundamental patterns in reality. A social contract supporting a global NES project, in other words, can be viewed as an authoritative collection of governance aspirations or ideals somehow present in nature as an underlying field, pattern or symmetry (this view is explored in greater detail in the next section concerning natural law). Denying the possibility of this 'eternal' dimension from which norms diffuse may be equivalent to concluding we must think in time and space wholly as a result of experience (a fallacy about the nature of thought, space and time Kant exposed in his *Critique of Pure Reason* thereby assisting to create the new physics of Einstein's age).

To distinguish these positions consider the social facts of *Magna Carta* (the English Barons' civil liberties contract with their King) or the constitutions of America and France (peoples' contracts with their political, military and judicial leaders, including constitutional texts like the *Virginia Declaration of Rights 1776*, the *American Declaration of Independence 1776* and the French *Déclaration des Droits de l'Homme et du Citoyen 1789*). The standard view is that such constitutions were an outcome of the popularisation of an intellectual ideal of a social contract created in philosophical writings such as those of John Locke, Thomas Hobbes and Jean Jacques Rousseau. Such constitutional arrangements might simultaneously exist, however, because of the gradual emergence in human understanding of a reality present in nature (though not necessarily correlating with common experience).

Civil and political, as well as economic, social and cultural rights are key components of such constitutional manifestations of a social contract. It is important to re-emphasise, however, the general absence from these documents of rights concerning how humans should use technology, or treat their environment or the other species that inhabit it.

Claiming a constitutional right has become a widely accepted philosophical and jurisprudential means of justifying one's action or inaction either morally or legally in the face of potential opposition. A large part of the popular attraction of such rights is that they can readily be asserted to act as 'trumps' over state policies and legislation. Rights, generally speaking, are effective insofar as they create enforceable duties in other persons. Rights claims inevitably exist in the context of those made by others and involve networks of mutual recognition that have a distinct symmetry when mapped geographically, temporally or in terms of spheres of influence.

In the late twentieth century John Rawls in his *Theory of Justice* influentially argued how the basic fact behind the social contract idea is that a society's principles, laws and rights emerge from general respect for a foundational virtue: justice. Rawls made sure to clarify that such a foundational understanding did not emphasise the good or goods aimed for by a society (peace, prosperity, and living room for example) above the individual civil and political rights and freedoms necessary to be preserved while such ends were achieved.

If asked to consider how a global NES project might fit within a social contract, Rawls would likely argue that as a species we should not attempt to flourish in life by seeking to achieve some aim or good independently defined, even if it is as ostensibly valuable as environmental sustainability. On such a view, humanity's goals and aims (equality in material conditions, environmental sustainability or expansion of consciousness, for example) are equally important with ensuring we consistently apply principles flowing from the virtue of justice to governing the conditions under which such goals and aims are formed and pursued. In the twentieth century, many regimes dominated by the social goals of communism (in relation to internal governance) as well as free-market capitalism (particularly concerning foreign relations) learnt this lesson slowly.

Amartya Sen makes the same point when he argues (pertinently in the present context) for technological and industrial development *with* freedom. In a free society, in other words, the free will and intrinsic dignity of each member should be valued prior to any ends collectively affirmed. To adopt a more Kantian mode of expression, it is the extent to which we consistently affirm principles in making governance choices that defines whether our society shapes its laws around virtues such as justice, equality and environmental sustainability or not. Kant thus considered practical reason supported 'freedom (independence from pressure exerted through another's discretion), insofar as it can coexist with every other person's freedom in accordance with a universal law'. Collective ends such as security, prosperity, economic growth or the free market cannot provide definitive characterisations of how laws should operate in a just society.

Ronald Dworkin's *Law's Empire* is another influential jurisprudential work that emphasises the foundational importance to a social contract of normative links between a society's respect for foundational virtues focused on individual legal rights and the achievement of specific national aims and goals. Dworkin views the idea of a social contract as uniting ideals that the tradition of Utopian political theory has gradually percolated into mainstream politics. Such ideals include the necessity for a fair political structure, a just distribution of resources and opportunities, and an equitable process of enforcing the requisite rules and regulations. Dworkin terms these the virtues of fairness, justice and procedural due process. They are human-centred, and environmental sustainability has no clearly defined place amongst them.

In societies run according to the rule of law rather than that of unbridled power, many citizens, as well as being able to anticipate how social force (police and military) will be used, are likely to feel that their association is regulated towards some optimal state that maximises the balance between individual free will and communal peace and security. Such political aspirations and hopes as we witness in the street demonstrations to remove dictators, or the individual acts of conscience to oppose them are often motivated by a sense that human conduct, to feel 'right', or 'natural' or 'complete', must conform to certain altruistic principles that form criteria by which statutes, judicial cases or arbitrary exercises of political power are criticised.

Respect for individual conscience (as a means of detecting and affirming such connectedness) is frequently one of the most important basic norms to emerge from a social contract. Conscience, as Emmanuel Levinas pertinently observed, is a potentially revelatory encountering of inner resistance to the thought or experience (direct or indirect) of arbitrary, violent or selfish use of freedom of will in denial of empathy with others. Levinas considered conscience was basically a 'conceptless' experience, but this can't be correct. Conscience has intuitive elements undoubtedly – once we explore it by regularly acting upon its promptings we are getting in touch with an aspect of ourselves that seems to exist in eternity or in dimensions of reality not confined by the three of space and one of time we are used to.

But far from being 'conceptless', conscience is made more subtle and responsive by education, even expanded through reasoning about the principles (such as those in the social contract) that we should apply in particular circumstances. There is certainly sociological evidence that many people feel themselves motivated by conscience. A crucial point is to consider the hypothesis that conscience exists in time and space the same way atoms or electromagnetic waves do. Physics has proven that photons and all matter exist simultaneously as particles and waves. Might conscience be the 'wave' part of us?

Evidence that environmental sustainability is becoming an issue of individual conscience may be found in the protests against use of new technologies in fishing, farming, industry, energy production and urban development that is deleterious to our environment. Such protests can be viewed as calls to renegotiate fundamental components of our global social contract so the laws emerging from it begin to support the foundational virtue of environmental sustainability.

Complex issues confront the potential use of a jurisprudential social contract as providing the governance framework for a global NES project.

One is that the influential academic expressions of the virtue-based social foundations of human rights according to Rawls and Dworkin are enmeshed in the US constitutional tradition and involve little, if any, express acknowledgement of public international law or the cultural, social and moral diversity that constitutes contemporary international civil society. Indeed it can be argued that a large part of the initial intellectual appeal of John Rawls' academic construction of a just social contract arose from the reluctance at that time of his nation, the United States, to embrace the great constitutive documents of the international human rights movement – for example the United Nations *International Covenant on Civil and Political Rights* and *International Covenant on Economic, Social and Cultural Rights.*

Another way of approaching the idea of a global social contract such as might underpin a global NES project is the notion that it represents a 'basic norm' or *grundnorm* (a type of a priori assumption) justifying the acts of free will by which individual humans agree to be governed or restricted by principles or laws. This concept has been held by some scholars of jurisprudence (notably Hans Kelsen) to support the proposition that a properly functioning legal system should aspire to be more than merely a concatenation of legislative and judicial facts undergirded by overt and covert political power. In making the case for a *grundnorm* at the conceptual heart of free society Kelsen arguably followed Kant in positing that minds characteristically human are preconditioned to operate not just in a field of three-dimensional space and time ('the starry sky above'), but of moral and legal norms ('the moral law within') to which governance systems must cohere ('become symmetric' we might say) if they are to retain their legitimacy.

Examples of a *grundnorm* might include the proposition that society should be governed by law, or that a society's laws should be true to its basic social contract. Other instances might be that 'the members of global society have a duty to co-operate with one another', that 'no member of global society should be deprived of life, liberty or property except by operation of legal principle justly applied', or that 'the members of global society have a duty to ensure a minimum level of health care, education, shelter, food, water and medicine to all its members, including through the equitable development and transfer of new technologies'.

Framed in such terms, a basic norm derived from the foundational social virtue underpinning a global NES project might become: 'the members of global society have a primary duty to ensure the sustainability of the natural environment including by the use of new technologies such as nanotechnology'.

The path to including environmental sustainability in our social contract theorising has been a long one. The practical culmination of anthropogenic (or human-centred) social contract reasoning on the international stage used to be viewed as the United Nations' *Universal Declaration on Human Rights* of 1948. This was particularly true of provisions such as Article 1:

> All beings are born free and equal in dignity and rights. They are endowed with reason and conscience and should act towards one another in a spirit of brotherhood.

Article 38(1) of the *Statute of the International Court of Justice* provides a guidepost as to the components of international law in a putative global social contract. It is useful to consult this article to determine what practical international legal manifestations may support a global NES project. Article 38(1) states that we are entitled to examine the contents of the following documents to discover what foundational virtues, principles and rights the international community has agreed upon:

a. international conventions, whether general or particular, establishing rules expressly recognized by the contesting states;
b. international custom, as evidence of a general practice accepted as law;
c. the general principles of law recognized by civilized nations;
d. judicial decisions and the teachings of the most highly qualified publicists of the various nations, as subsidiary means for the determination of rules of law.

One way of viewing the task of proposing nanotechnology-based environmental sustainability as a primary component of a global social contract would thus involve consideration of international legal documents falling under the above subheadings. United Nations Declarations and Conventions are likely to be the most prominent instruments involved.

Strangely, however, the idea of a global social contract (one in which all the peoples of the world are taken to have agreed upon their basic governance arrangements), or that core United Nations Declarations and Conventions such as the *Universal Declaration on Human Rights* might be practical expressions of it, has never been a point of consensus in jurisprudential circles. A reason for this may be that the notion of such a global social contract cuts across the interests of the cliques of political, business and military leaders who control nation states. Claims of jurisprudential evolution towards justiciable and enforceable international human rights as part of any functional global social contract are still viewed by many academic thought-leaders (particularly those less well informed about the political and legal power of corporate globalisation) as implying a problematic combination of naïve assurance about the regulatory importance of ideals and mistrust of governments to provide peace and security.

In international law, the concept of sustainability wends back to areas such as forestry and fisheries – 'sustainability' meant the maintenance of the long-term productivity of these resources. The *Convention on Wetlands of International Importance* (*Ramsar Convention*) uses the phrase 'wise use' instead of sustainability. Sustainable utilisation thus was defined as 'the human use of wetland so that it may yield the greatest continuous benefit to present generations while maintaining its potential to meet the needs and aspirations of future generations'. The *Brundtland Report* by the United Nations Commission on Environment and Development defined sustainable development as 'meeting the needs of the present without compromising the ability of the future generations to meet their own needs'.

Principle 1 of the *Rio Declaration* states that 'human beings are at the centre of concerns for sustainable development [and] they are entitled to a healthy and productive life in harmony with nature'. The *Stockholm Declaration* affirms that any '[hu]man has the fundamental right to freedom, equality and adequate conditions of life, in an environment of quality that permits a life of dignity'. The World Commission on Forests and Sustainable Development likewise explicitly states that the whole issue of forest ecosystems *is about people*. The report compiles the number of lives and livelihoods lost as a result of disruption in the forest ecosystem.

The UNESCO *Declaration on the Responsibilities of the Present Generations Towards Future Generations* is an agreement by nation states on ethical principles directed to aspects of sustainability that may potentially become part of international customary law if enough states obey it with the intention that it creates legal obligations. Amongst the Declaration's provisions is the following, drafted in the sincere but somewhat clumsy language of instruments prepared by international committee:

> The present generations have the responsibility to bequeath to future generations an Earth which will not one day be irreversibly damaged by human activity. Each generation inheriting the Earth temporarily should take care to use natural resources reasonably and ensure that life is not prejudiced by harmful modifications of the ecosystems and that scientific and technological progress in all fields does not harm life on Earth.

Judge Weeramantry of the International Court of Justice has stated that the principle of sustainable development as formulated in the *Brundtland Report*, the *Rio Declaration* and United Nations *Agenda 21* forms part of the basic principle of law of civilised nations under Article 38(1) of the *Statute of the International Court of Justice*. He stated that damage to the environment can impair and undermine all the human rights spoken of in the United Nations *Universal Declaration on Human Rights* and other human rights instruments. Thus, he held, while governments have the right to initiate development projects for the benefit of their people, there is likewise a duty to ensure that those projects do not significantly damage the environment.

The judge found the principle of sustainable development supported by resolutions of the United Nations General Assembly, and various United Nations conventions (treaties binding under international law) such as the United Nations *Convention to Combat Desertification in those Countries Experiencing Serious Droughts and or Desertification, Particularly in Africa)*, the United Nations *Framework Convention on Climate Change*, and the *Convention on Biological Diversity* (CBD). The CBD for the purpose of creating international legal norms defines an ecosystem as a 'dynamic complex of plant, animal and micro-organism communities and their non-living environment interacting as a functional unit'.

Also supporting the norm of sustainable development were the preamble of the *North American Free Trade Agreement* between Canada, Mexico and the United States, the preamble to the *Marrakesh Agreement* of 1994, establishing the World Trade Organization (WTO), as well as various key statements of the World Bank Group, the Asian Development Bank, the African Development Bank, the Inter-American Development Bank, and the European Bank for Reconstruction and Development.

Judge Weeramantry's dissenting opinion is far from mainstream jurisprudence in speaking of a duty upon all members of international civil society to preserve the integrity and purity of the environment. He stated that when Article 38(1) of the *Statute of the International Court of Justice* described the sources of international law as including the 'general principles of law recognized by civilized nations', it expressly opened the way for entry into that corpus of principles from traditional legal systems – 'embodying as they do the wisdom which enabled the works of man to function for centuries and millennia in a stable relationship with the principles of the environment'.

Christina Voigt, in her book *Sustainable Development as a Principle of International Law*, likewise has argued that obligations of environmental sustainability have now attained the status of general principles of international law under Article 38(1). Yet, as mentioned, even this idea of sustainability under international law is oriented towards the needs of the world's human population. The preservation and regeneration of ecological systems and resources are encompassed within it only to the extent that they sustain human economic development.

These various definitions suggest two dimensions to the notion of sustainability as it might fit within the international law components of a global social contract supporting a macroscience NES project. First, there is a spatial dimension: intra-generational equity between humans of the same generation living in different places of the Earth. Second, there is a temporal dimension: future generations should be considered in the decision-making of the present.

Another problematic issue here is the connection between an NES project, a global social contract and the conception of 'rights'. As mentioned, claiming a 'right' has become a widely accepted jurisprudential means of justifying an individual's demands upon other citizens under social contract theory. A large part of the popular attraction of 'rights' is that they can readily be asserted to act as 'trumps' over state policies and legislation. Rights, however, can only reside in persons, in conscious entities capable of interests. Ecosystems do not fall into that category on traditional jurisprudential analysis.

Domestic courts nonetheless have shown a high level of ingenuity in extending the human rights protections enshrined in their national constitutions to cover concerns about sustainability. In the Indian case of *Rural Litigation & Entitlement Kendra* v. *Uttar Pradesh*, the Court issued an order to cease mining operations, stating that the petitioners had the right to live in a healthy environment. In the South African case of *Minister of Health & Welfare* v. *Woodcarb Pty Ltd*, the Court upheld a citizen's right to an environment that is not detrimental to health under the South African Constitution – the Court concluded that the defendant's unlicensed emissions illegally interfered with the neighbours' constitutional right to a healthy environment. In the Tanzanian case of *Festo Balegele* v. *Dar es Salaam City Council*, the court ordered the defendants to cease dumping garbage in the area and to construct a dumping ground where such dumping would not be a threat to the health of the residents.

Could international human rights law provide jurisprudential support for a global NES project? While international environmental law is a *relatively* new area of law, the international law of human rights is a well-established legal regime, which is almost as old as the *United Nations Charter* itself. The United Nations *International Covenant on Economic, Social and Cultural Rights* (ICESCR) has attained almost universal membership by ratifying nation states. It is unambiguously part and parcel of international law.

Article 12 of ICESCR proclaims the 'right of everyone to the enjoyment of the highest attainable standard of physical and mental health'. There are countless examples of threats to human health caused by the disruption of the sustainability of ecosystems, for example the contamination of water and food supply, and air pollution leading to acute respiratory diseases. In its General Comment on the Right to Health, the *Committee on Economic, Social and Cultural Rights*, which was set up by the ICESCR, makes it clear that it is 'incumbent on States parties and other actors in a position to assist, to provide "international assistance and cooperation, especially economic and technical" which enable developing countries to fulfil their core and other obligations'. Violations of the right to health can also occur through the omission or failure of states to take necessary measures arising from legal obligations, including the obligation to provide technological assistance.

Though not considered part of international law, the United Nations *Millennium Development Goals* also may be viewed as encapsulating core components of an emerging global social contract that might support an NES project. These goals require nation states to:

Goal 1: Eradicate extreme poverty and hunger

Goal 2: Achieve universal primary education

Goal 3: Promote gender equality and empower women

Goal 4: Reduce child mortality

Goal 5: Improve maternal health

Goal 6: Combat HIV/AIDS, malaria and other diseases

Goal 7: Ensure environmental sustainability (emphasis added)

Goal 8: Develop a Global Partnership for Development

Goals 7 and 8 above particularly could support nanotechnology-based environmental sustainability as a primary global social virtue.

The UNESCO *Universal Declaration on Bioethics and Human Rights* (UDBHR) (also not yet a fully fledged part of international law under Article 38(1) of the ICJ Statute) likewise includes the environment beside more traditional global public health concerns in what can be viewed as an attempt to scope basic principles of a global social contract. Article 14(2) of the UDBHR provides:

2. Taking into account that the enjoyment of the highest attainable standard of health is one of the fundamental rights of every human being without distinction of race, religion, political belief and economic, or social condition, ***progress in science and technology should advance***:

(a) access to quality health care and essential medicines, especially for the health of women and children, because health is essential to life itself and must be considered to be a social and human good.

(b) access to adequate nutrition and water;

(c) *improvement of living conditions and the environment*;

(d) elimination of the marginalization and the exclusion of persons on the basis of any grounds

(e) reduction of poverty and illiteracy. (emphasis added)

Another 'soft law' expression of norms that might support a global NES project is found in the *Earth Charter* (2000): 'we must join together to bring forth a sustainable global society founded on respect for nature, universal human rights, economic justice, and a culture of peace. Towards this end, it is imperative that we, the peoples of Earth, declare our responsibility to one another, to the greater community of life, and to future generations.'

Technology transfer is a further well-established area of international law that putatively supports the pace of a macroscience NES project in a global social contract. Technology transfer will be discussed in detail (in relation to different options for a global NES project) in subsequent chapters. The argument here is that hypothetical founders of any social arrangement placed behind a putative 'veil of ignorance' to shape the principles and rules that would thereafter govern their global relations would surely have taken into account how our new technological advances should advance the interests of species survival and integrity of ecosystems.

One way in which social contract analysis may be viewed as considering how human beings should use technology in relation to their environment involves the precautionary principle. This emerged in German regulatory policy during the 1970s and rapidly spread through the international policy arena.

The precautionary principle has now been incorporated into a number of specific laws on environmental protection relating to the ozone layer, climate change, biological diversity, fisheries management, and extending even to the protection of human health generally notably including food safety. The development and application of the precautionary principle has never been free from criticisms. Some commentators argue, for example, that a precautionary measure may have adverse effects, rendering protective measures hazardous in themselves to the environment or human body. Others warn that it may well result in preventing development of new technologies that may serve to alleviate the challenged environmental harm.

The precautionary principle was a philosophical challenge against policies and laws that demanded an often unrealistic level of scientific certainty about risks of marketing or otherwise introducing new technologies before recommending or implementing public health and environment protection measures. One well-known international enunciation of the precautionary principle is found in Principle 15 of the *Rio Declaration on Environment and Development*: 'Where there are threats of serious or irreversible damage, lack of full scientific certainty shall not be used as a reason for postponing cost effective measures to prevent environmental degradation.' My own publications have explored the precautionary principle particularly in relation to regulation of nanoparticles in sunscreens (especially when applied to damaged skin), nanosilver in waste water and carbon nanotubes in settings where they can be inhaled.

In summary, this section has attempted to set out how the notion of a global social contract based on the rule of law might provide a workable jurisprudential foundation for a global NES project. In that context we have examined how social virtues give rise to and are supported by legal principles and rights, including constitutional and international environmental and human rights law.

In the next section we consider in greater detail whether the foundational social virtue underpinning a global NES project may represent a 'law' of nature that is just as 'true' as the equations of thermodynamics, quantum mechanics, general relativity or (as may soon be established experimentally as well as mathematically) string theory.

2.3 NES AND NATURAL LAW THEORY

After my first year of legal studies I spent a few months alone walking and catching trains through Europe in winter. I kept a diary and experimented with following my intuition. One of the most intriguing places I visited (through an interesting series of coincidences) was the Findhorn community near Forres in northern Scotland. Although I didn't realise this till later, the Findhorn community was already well known for pioneering a vision of humans living in harmony with nature. While there I first came across the Japanese haiku poet Basho's *Narrow Road to the Deep North*, which to me became an instructive meditation on the fragility of the human journey through life.

About this time I also became interested in the music of the Breton musician Alan Stivell. Stivell's album *Renaissance of the Celtic Harp* contains a track entitled 'Ys', which evokes the inundation of an ancient Celtic harbour city by rising sea levels, supposedly as a punishment for how its leaders had disregarded the laws of nature. I became interested in the widespread cultural importance accorded to similar apocalyptic myths that appeared designed to prevent disharmony with principles underpinning the natural world.

I began to wonder whether scholars of governance and physics made notable advances (such as the systems of international human rights and equations of quantum mechanics respectively) by a similar process – seeking to reflect a harmony and geometric symmetry presumed to exist in nature. This section develops arguments justifying why this type of approach to governance arrangements may be termed 'natural law' and aims to provide jurisprudential support for a global NES project.

Natural law theorists posit that there are certain inherent ideal principles (or innately present deontological axioms) of human conduct, capable of discovery by human reason, with which man-made law should conform. They have hitherto tended to rely on the recorded perspicacity of religious founders (particularly Jesus) or eminent scholars (for instance St Thomas Aquinas) for insights about these principles. Traditional expressions of the natural law approach thus claim that the laws of the natural world are best viewed through the prism of a particular religion to provide the template for the laws that should regulate human society. This tends to make acceptability of natural law conditional upon allegiance to the religious ideology that underpins it.

As the eminent scholar HLA Hart pointed out, natural law is often characterised jurisprudentially as one of many ways to think about the relation between concepts of morals (personal or community knowledge of right and wrong conduct) and law. Scholars who support the governance approach of legal positivism adopt the position that there is *no necessary* connection between these ideas. According to legal positivists, law is simply what socially accepted rules of recognition (such as those in a constitution, legislation, treaties or judicial decisions) say it is. In other words, under the so-called 'positivist' approach the absolute and fundamental fact of legal obligation comprises the provisions in a national constitution or international treaty that specify, for example, how governments and states are recognised, how statutes should be passed by elected representatives in their respective parliaments, the requirements for judges to be duly appointed and their pronouncements officially recognised, or what the International Court of Justice can consider to be international law.

The 'positivist' conception of social law undoubtedly holds sway today amongst the senior minds in global academia, policy and governance. It supports the view that legal systems (such as those regulating a nanotechnology-embedded world premised on environmental sustainability) will fundamentally comprise legal documents (constitutional provisions, international treaties and custom, statutes and judicial decisions, contracts, regulations, professional and institutional guidelines) themselves composed of basic units called norms.

Such basic norms, the positivist view holds, emerge as sociological facts, not as manifestations of ideals cohering with some harmony underlying reality. Varied interpretations may be given to such norms (for example depending on the cultural mores and legal training). Nevertheless, legal positivists hold, no matter how much some citizens might wish to protest against or refuse to obey a law, if that law has satisfied the appropriate constitutional 'rule of recognition' there can never be a point where it is so unjust as to no longer be capable of being called a law. It might be a 'bad' law, but it is still a 'law'.

At the heart of the 'natural law' enterprise, however, remains the intriguing notion that governance systems should reflect the search for and application of universally applicable ideals rather than convenient social compromise. Thus, the 'natural law' position in any of its formulations supports the position that this world includes some suprasensible component – resembling a consciousness of the universe – in harmony with the laws of the physical world and with which mature humans seek to make their lives coherent.

Natural law theorists are frequently derided by such so-called 'blackletter' positivist lawyers as falling for the fallacy implicit in different meanings of words such as 'must', 'bound to', 'ought' and 'should'. Physical laws, the positivist argument runs, like those of thermodynamics, quantum mechanics and general relativity, are descriptive outcomes of observation, experiment and reasoning. They are qualitatively different from prescriptive laws regulating human behaviour. Such critics often express fears about the capacity of totalitarian ideologies to prey upon loyalty to natural law's mystical conceptions of governance and despair at the lack of practical law enforcement that seems implicit in them.

Natural law theorists have been hard-pressed to effectively answer such criticisms without appeals to religious faith, in large part because the concept of natural law remains so closely linked to the view that the observed regularities of nature arise from an anthropomorphised Divine Legislator of the universe. Scholars prominent in the Roman Catholic religious tradition have been particularly supportive of natural law theory, in part because of their ideological claim (contested by the Protestant churches) that the Pontiff in Rome (rather than a congregation of the faithful individually informed by fellowship, scripture and conscience) is capable of infallibly expressing the will of the Christian God on Earth.

Some natural law theorists, not wishing to rely entirely upon religious doctrine for relevant insights, have examined the literary records of intellectual controversies in metaphysics, theology, anthropology, sociology, political science and history to find traces of what laws 'ought' to prescribe for us. This type of process, however, tends to represent merely a historical investigation of prior opinions about natural law and related areas of inquiry. In the contemporary world of conflicting religious fundamentalisms, increasingly prevalent atheism and politically soporific consumerism, religious-based natural law is unlikely to provide a fruitful ground for governance models supporting a global NES project.

Governance mechanisms already exist whereby natural law conceptions, if appropriately framed, can readily be developed into positivist norms. The English common law is one example of such a system. Pertinent examples here are incremental judicial evolution of tortious liability in damages for not only negligent acts done to a 'neighbour' with whom you have no contract, but to particularly vulnerable people by people with a power advantage over them through pre-existing relationships. Judges explaining such decisions often do so by making reference to their attempts to achieve coherence with certain fundamental principles that should be present in human relationships. Civil and political human rights also have been a major vehicle whereby positivist norms may be viewed as developing out of natural law type reasoning.

Under international law, natural law conceptions have been described as underpinning the principle of *Ius Cogens* (as expressed in Article 53 of the *Vienna Convention on the Law of Treaties*). The mechanism for enunciating *Ius Cogens* norms, otherwise described as pre-emptory or non-derogable norms of international law, has been described as involving the positivisation of natural law. Commentary by academics and judges of the International Court of Justice describe such *Ius Cogens* norms as derived from moral intuitions, the inner moral aspiration of the law, the common moral goals of the community of nation states and as involving a normative category with an open-ended character. Characteristic core examples include prohibitions on torture, slavery and genocide. Government prohibitions or systematic elimination of the rule of law, or any respect for human rights (by for instance banning police or courts) are possible inclusions. So too could be attempts by a nation state to systematically destroy the ecosystem of its people or that of the planet.

Another potential example of the emergence of natural law principles into the positivist arena involves the jurisdiction of the International Criminal Court (ICC) over crimes against humanity. Such crimes and the jurisdiction of the ICC to enforce them appeal on one analysis to a set of basic norms at the heart of all domestic legal regimes. A further instance relates to the provisions in United Nations Declarations and Conventions that certain areas (outer space, the moon, the deep sea bed, Antarctica, the human genome, the world's cultural and natural heritage) should be considered 'common heritage of humanity' with attached obligations on nation states to prevent military use, private exploitation and ensure universal use.

It has been a standard part of natural law theory to say that such principles emerge from the sociological fact of global human ethics, whether or not influenced by religious tradition. It will be argued here, however, that natural law philosophers are just as capable of turning to physical science to discern the ideals and patterns of symmetry that society should reflect. Indeed this change may facilitate environmental sustainability being incorporated into natural law theory. Science-based natural law as an academic discipline could take up challenges such as forging links between governance systems and those of physics; establishing that the world's great religions do not exhaust conceptions of the spiritual or its importance to contemporary scientific and political decision-making; countering the de-sacralisation and disenchantment of our relationship with the environment as a prerequisite to commodifying and exploiting it.

Science-based natural law could explore, for example, whether our conscience responds to patterns of symmetry in events, deviations from which our reason portrays as injustice or inequity. It could consider what physical laws (including notions of indeterminacy) support the processes whereby we strive to temper our egocentricity by minimising harm to an ever-widening circle of life in nature. Much of the theory's value will reside in its power to predict what will happen as governance systems strive for symmetry with those of the natural world, helping to achieve a more complex and accurate understanding of the necessary synergies between our laws and those of the environment.

That our fundamental understanding of societal law should be so informed by scientific research is not a new idea. We see this not merely in use of scientific evidence to determine standards of care for liability in non-contractual civil wrongs, or in contrasting theories of causation, but in the gradual evolution of rules and principles organising how we should relate with each other, other species and our environment. Public health legislation (for instance specifying standards for clean air, water and food as well as access to safe medical services and medicines) and statutes to abolish slavery, grant the vote to women, reduce crime, protect animals and reduce pollution, as well as treaties (like those on nuclear and biologic arms control, or designed to prevent torture and genocide), were in considerable measure driven by increased scientific information operating synergistically with an expanding public conscience. What would be different, however, in a science-based natural law approach (such as might underpin governance of a global NES project) is the use of science as a source of ideals towards which the law should aspire.

Academic discourse tends to markedly distinguish the process of under-standing and applying physical laws (such as those governing the inter-action of forces and matter at or below the nanoscale, or the propagation of energy and matter through the universe) from laws passed by parliaments and interpreted by judges. Further, it often appears to accept as an unques-tionable truth the fact that the two should bear little sensible relation to each other in both process and outcome.

Yet the idea that scientific advances should provide insights for the development of governance concepts (including legal principles and rules) is not a recent or unduly peculiar one. A notable example from as far back as the 1600s was Thomas Hobbes' *Leviathan,* which sought to apply Galileo's concepts of gravitational forces to constitutional arrangements (self-interest, Hobbes considered after meeting with Galileo in Florence, will keep the people of any nation moving in a straight line till met with an equal and opposite force). In the nineteenth and twentieth centuries, jurisprudential scholars likewise attempted to apply Charles Darwin's theory of evolution and, later, concepts of statistical mechanics to legal arrangements. Though not without controversy, these approaches led to reforms still considered valuable against a range of different criteria.

The argument here is not that governance principles supporting nanotechnology-based environmental sustainability might derive support from neuroscience research (for example confirming altruistic ideals in our cerebral control centres (like the hypothalamus and limbic system) along-side hate, love, guilt and fear and other factors that create our sense of 'good' and 'evil'). Lawyers are used to considering such information in relation to areas such as criminal responsibility. The argument, rather, is that the scientific disciplines that could provide the basic insights for natural law theory range much more widely – to include those related to the origin of matter (baryogenesis) and origin of life (biogenesis), as well as thermodynamics, the Standard Model of particle physics and Einstein's general theory of relativity. Further, this approach opens the door to governance systems embracing idealistic positions that accord with science, but not sociological facts.

To reiterate, in his *Critique of Pure Reason,* Kant stated our common sense can as little consider simultaneous times (of more than one dimen-sion) as it can successive spaces (of more than three dimensions). An extension of such reasoning (picked up by Albert Einstein who studied Kant's *Critique of Pure Reason* in his youth) was that lack of correlation with everyday sensory experience should not be considered a conclusive argument against the existence of physical laws that otherwise reveal a geometric and mathematical harmony in nature.

The application to jurisprudence of this insight (that reality may be underpinned by harmonious laws not necessarily confirmed by sensory experience) hitherto, as mentioned, has been the task of natural law theorists chiefly from the Christian religious tradition. They did not approach the issue of social law (as Kant did) by means of analogy with the physical sciences, but rather through the lens of established religious dogma.

The value of such an approach may become plainer when we consider modern physics in more detail as a source (even by analogy) of natural law norms that might underpin a global NES project. As late as the end of the nineteenth century, *atoms* were generally regarded as symbolic entities, much as natural law jurisprudential *norms* or even conscience are today by those with a positive law perspective. Then in 1908 Jean Perrin demonstrated the mechanical effects (in Brownian motion) of what were deduced to be tiny particles. In 1909 Ernest Rutherford and his experimental team streamed radium-emitted alpha particles (more massive than electrons) into the atomic nucleus of zinc sulphide. Once in 8,000 attempts, they observed by use of a gold foil scintillation screen that such alpha particles would bounce back as if off something very hard. They deduced that, although most of the atom was what they considered to be empty space and its outer limits were only roughly demarked by the movements of its electrons, at the core of the atom, and thus at the heart of matter, was a tiny nucleus with mass of extraordinary density. Rutherford further established that if 300,000 volts were used to accelerate a proton into that nucleus, the latter emitted alpha particles. The atom, an entity that had been a hypothesis, a component of idealised reality like environmental sustainability is now in many legal circles, was proven to exist.

Increasing knowledge of the atomic basis of reality facilitated the nanotechnology revolution and provides many potential analogies for the development of social norms. It has now, for example, been established that an atom containing an equal number of protons and electrons is electrically neutral. If this balance is unequal, the atom has a positive or negative charge and is called an ion. The number of protons in an atom's nucleus determines its chemical element, and the number of neutrons in the nucleus determines its isotope. The radius of an atom also varies with its type of chemical bond, the number of neighbouring atoms (coordination number) and a quantum mechanical property known as spin. Can such knowledge provide insights about how governance norms should operate to balance opposing social forces if they are being used to achieve environmental sustainability?

More peculiar features of physics may be drawn upon by natural law in this context. Atoms and photons, for example, have particle and wave properties that cannot be detected by the same observational technique simultaneously. In 1964 John Bell devised a testable prediction (now known as Bell's inequality) based on two reasonable assumptions: that the measurement of one particle cannot instantaneously influence another, distant particle (locality) and that particles have properties before you measure them (reality). Numerous experiments (such as that of Einstein, Podolsky and Rosen) have since proven that the universe constantly violates Bell's inequality. Are there also valuable perspectives to be gained here about how social laws such as those supporting a global NES project should work?

Physicists themselves gained many of the theories that, after experimental verification, led to such knowledge of atomic structures, by reasoning by analogy from imaginative ideals. Thus, the theory that electrons orbited the atom was developed by analogy from observations of planets orbiting the sun. When Niels Bohr, for example, heard of Ernest Rutherford's model of an atom with a positively charged nucleus surrounded by a negatively charged electron or electrons, he drew on Johannes Kepler's laws of planetary motion in trying to determine paths of the electrons. Similarly, once it was realised that the phenomena of interference and diffraction showed that light had a wave function, but that photo-electric emission and scattering by free electrons also revealed light's simultaneous particle composition, it was reasoning by analogy that applied such insights to *all* material particles.

Likewise, the physical 'law' that related all known elements in the periodic table to their atomic weight was intuited by Dmitri Mendeleev by analogy to principles of mathematical harmony in the universe long before its explanation was known in terms of atomic structure. The existence of the neutrino (a particle with neutral charge and very small mass) arose as a result of reasoning from an established principle of conservation of matter and symmetry. Reasoning by analogy also led to the discovery of quarks composing particles (hadrons such as protons) that interact with the strong force that binds atomic nuclei, as well as proof of leptons (such as electrons) that are not affected by the strong force. The use of analogy is also implicit in the search for the Higgs boson and the nature of the force that hides the symmetry between the weak and electromagnetic interactions, as well as the multiple extra dimensions that may be 'curled up' according to string theory in our familiar three dimensions of space and one of time.

Some well-known physicists, notably Wolfgang Pauli, have been critical of this mode of reasoning scientific principles from analogies-based appreciation of natural harmony and symmetry. Yet Pauli also declared his support for reasoning that might link the search for social and physical laws. 'The process of understanding nature' he claimed 'as well as the happiness that man feels in understanding, that is, in the unconscious realization of new knowledge, seems thus to be based on a correspondence, a matching of inner images pre-existent in the human psyche with external objects and their behaviour.' This is the type of synergy between human consciousness and fundamental physics that may be reflected in science-based nature law, particularly as it shapes norms associated with environmental sustainability.

The laws of thermodynamics provide another potential source of analogy for science-based natural law norms that might support a global NES project. The first law of thermodynamics holds that when energy is transferred from one system to another it is conserved and transformed rather than created or extinguished. The second law of thermodynamics holds that in any isolated system (not exchanging matter or energy with others and including the universe as a whole), there is an irreversible tendency (a 'time's arrow') of gradients in physical processes such as temperature, pressure and chemical potential to disappear. This tendency to 'running down' is called entropy and the highest entropy state is where all chemical elements in a system are equally dispersed and matter homogenised (the so-called 'heat-death' of the universe).

Economists such as Nicholas Georgescu-Roegen, Herman Daly and Robert Ayres have developed theories that the first and second laws of thermodynamics impose constraints on materials transformation processes that are essential to the economy. Natural law scholars have been less willing to explore thermodynamics as a template for social norms, though this could be very fertile intellectually, it is argued here, in relation to concepts like environmental sustainability.

As another illustrative example, science-based natural law could explore the potential correspondence between the intellectual method of reductionism as a means of revealing physical laws and the process of social law reform such as would be involved in establishing a global NES project. Thus, much nanoscience research is based on insights provided by use of particle accelerator or synchrotron magnetic fields to accelerate electrically charged particles as well as X-ray photons (to interact with the electron cloud surrounding the nucleus) and electrically neutral neutrons (to collide with the nucleus itself). This process of colliding particles with the components of the atomic world, being successfully used to reveal ever more intricate information about the nature of matter, could also provide insights relevant to locating environmental sustainability within a science-based form of natural law.

To all this, however, must be added the caveat enunciated by the physicist Paul Dirac in his classic text *The Principles of Quantum Mechanics*. Dirac doubted that one could ever determine the physical laws governing the ultimate structure of matter simply by reducing investigation of it to smaller and smaller parts. He pointed out in terms very pertinent to a science-based natural law approach to governance of nanotechnology-based environmental sustainability:

> So long as *big* and *small* are merely relative concepts, it is no help to explain the big in terms of the small. It is therefore necessary to modify classical ideas in such a way as to give an absolute meaning to size. At this stage it is important to remember that science is concerned only with observable things and that we can observe an object only by letting it interact with some outside influence. An act of observation is thus necessarily accompanied by some disturbance of the object observed. We may define an object to be big when the disturbance accompanying our observation of it may be neglected, and small when the disturbance cannot be neglected. This definition is in close agreement with the common meanings of big and small.

Science-based natural law theorists might make an analogy from this to the distinction between a 'big' law that remains uninfluenced by what Dirac terms the 'gentleness of observation'. Yet there must be a limit to the powers of observation to determine Natural Laws. What happens to atoms or social laws beyond that limit? As Dirac notes:

> In order to give absolute meaning to size, such as is required for any theory of the ultimate structure of matter, we have to assume that *there is a limit to the fineness of our powers of observation and the smallness of the accompanying disturbance – a limit which is inherent in the nature of things and can never be surpassed by improved technique or increased skill on the part of the observer.*

The consequence of such a limit for a science-based natural law theory is not that reality at a level beyond the basic norms of our society ceases to exist for us. Rather, it means that it is easier to apply laws to big objects where the disturbance involved in observation is negligible, than to do so in relation to the smallest objects – where the act of observation will significantly promote uncertainty of causation. As Dirac put it:

> Causality applies only to a system which is left undisturbed. If a system is small, we cannot expect to find any causal connexion between the results of our observations.

Perhaps the most distinct contemporary symbol for this reductionist and unification process to discovering basic physical laws is the Large Hadron Collider (LHC). This 27 km particle accelerator (the world's most powerful synchrotron) was the outcome of a global macroscience project, being built by the European Organization for Nuclear Research (CERN) near Geneva on the border between France and Switzerland. Physicists from around the world aim to use magnetic fields in the LHC to accelerate and collide electrically charged particle beams of either protons (7 trillion electronvolts (1.12 microjoules) per particle), or lead nuclei (574 TeV (92.0 μJ) per nucleus). They hope the results may explain such fundamental questions as how the concept of mass arises, what the undetectable 'dark matter' is that makes up such a large proportion of the universe and whether the equations of quantum mechanics and general relativity elegantly fit, as string theory predicts, in a universe with more dimensions than the three of space and one of time known in our current experience. Science-based natural law might go further and posit the eventual unification of all physical *and* social laws.

The beauty of a jurisprudence of science-based natural law, as with physics, is that we may be able to develop and discern their respective and interrelated principles without having a complete picture of how they all fit together. We developed the presumption of innocence and the rights of habeas corpus, for example, long before freedom of speech and association. Likewise, Newton came up with a workable theory of gravity well before we understood its attraction at a distance was owing to the capacity of matter to warp space and time. The social contract, as with the equations of quantum mechanics and general relativity, similarly was not required to emerge fully formed and immutable (though some conservative constitutional lawyers operate under the fallacy that such is how constitutions should be interpreted).

We have spent some time discussing the issue of whether nano-technology-based environmental sustainability can be fitted conceptually within a science-based natural law theory. One reason is that scholars of science-based natural law are likely to play an important role in achieving greater coherence between public deliberation, civic participation and concepts of environmental sustainability that will be important for the success of a global NES project. They will help frame the language and concepts of such debates including those involving existing international law concepts such as 'human dignity', 'sustainable development', 'inalienable rights', 'technology transfer', 'progressive development', 'proportionality' and the 'margin of appreciation'. Such a theory may become a powerful mechanism for exploring the political and economic tensions likely to underpin a world where nanotechnology is governed primarily to promote environmental sustainability.

Science-based natural law theorists also are likely to play an important role in the inevitable public debate over resource allocation in relation to a global NES project and related ethical questions such as whether and how environmental sustainability should become a major concern of our character, informing not only isolated acts of choice, but the whole course of our life.

2.4   GLOBAL SIGNIFICANCE OF LOCAL NANOTECHNOLOGY GOVERNANCE

I remember once walking with Justice Lionel Murphy outside the High Court at lunchtime past a group of men and women playing touch football on the grass. "You watch," he said, "if they keep doing this, in a few months there'll be laws regulating where they can play, how many of them can do it, how much they have to pay in insurance." Lionel Murphy knew the importance of laws allowing a balance between individual freedom and social good. As federal Attorney General he had passed some of the first environment protection legislation in Australia. He'd taken the French government to the International Court of Justice to protest against their continued atmospheric nuclear tests in the South Pacific. He'd raided the offices of Australia's secret service when it refused to provide him with necessary information. Murphy was someone who believed passionately in the role that law should play in shaping not only a just and equitable but also a sustainable society.

If we look at a building, a park, a garbage tip, people walking, talking, loving, fighting or begging, it is increasingly true that all these contemporary objects and activities exist in a fundamental sense because of both atoms and legal norms. Today it is not just atoms, but laws that determine whether societal infrastructure is safe, whether different species of trees or plants or animals exist and in what location, if people can use different forms of technology as they wish, or go hungry or become obese, whether banks can cause financial instability on a global scale or whether corporations exclude sick people from the health insurance coverage they paid for, or pollute and destroy ecosystems.

As a result of the rapid growth of the human population, as well as the technological processes and natural resources exploited to satisfy its needs, there will soon be almost no activity on the planet that continues to function unless some legal norm permits it to. Whether species of plants or animals survive or become extinct, whether rain or snow falls or does not, whether forests grow or are chopped down, whether cities beautify or are destroyed, will soon be as much a function of the legal norms that fundamentally regulate their existence as their atomic constituents. The standard definition of a species, for example, was based on the capacity of its members under natural conditions to mate and produce fertile offspring.

Human knowledge of and capacity to manipulate genetics, as well as loss of biodiversity, now means legal norms, in many cases, will be the crucial factor in determining whether breeding continues. We may even go further and suggest that one day all the physical laws in the universe (including the unidirectionality of time in the dimensions we are familiar with) may arise because of some form of legislation, of a conscious will that it be so.

It is trite to say that technological advancement, of which nanoscience is merely one of the recently most advanced phases, offers humanity unprecedented power over the natural world, without necessarily enhancing the wisdom needed to use it well. Paradoxically the real question in terms of our survival and moral advancement, it is increasingly being realised by scientists, ethicists and lawyers, is learning to not think only about ourselves. The Human Genome Project, for example, when set alongside the critical contemporary problems of global poverty, biodiversity loss and environmental degradation risks being viewed historically as an exercise in scientific narcissism.

I often tell my students that their career choice is not a matter of thinking what they 'want' to do, but what they 'should' do – what duty is it likely the universe will conspire to help them achieve. Likewise, it is argued here that scientists and policy-makers must undertake an act of collective altruism and earnestly consider 'what is our critical duty in relation to our environment, for our mutual benefit in this period of great peril?' Answering this requires the human species to garner its capacity to innovate new technology to new global governance arrangements – that treat our collective existence as of equally moral worth with that of other life forms. Encouraging this type of thinking is necessary for widespread public support of environmental sustainability as the primary social virtue of the coming nano-age and the governance foundation of a global NES project.

Most humans, regardless of their religious persuasion or lack of it, accept science has proven that atoms are a basic component of this world. We've seen how our knowledge of atoms developed from analogies (for example of electron shells to the solar system) as well as the idea that truths about the universe may not necessarily correlate with sensory experience or the basic mode of reasoning we term common sense. This chapter has investigated the proposition that drawing analogies from such scientific understandings of how the world is fundamentally composed must figure more prominently in governance theory if we are to draw the best out of nanotechnology and live sustainably in this world.

In other words, we must consider the hypothesis (however much it appears not to correlate with sensory experience) that individual and social virtues, such as justice, fairness, respect for human dignity or ecological sustainability (as well as our resonance with them in conscience) cannot be completely defined as human inventions; instead like atomic structure, geometry, mathematics, time and consciousness, they may be harmonics (albeit imperfectly discerned) of natural laws – inherent in the foundational structures of the cosmos in which we exist.

The following chapter explores the extent to which social laws not emerging from a fundamental moral and jurisprudential understanding prioritising the virtue of ecological sustainability (be it associated with a social contract or science-based natural law) may create obstacles for a global NES project. This book accepts that few scientists involved in nanotechnology research will have a detailed understanding of international legal norms associated with public health or environmental sustainability. For many, no doubt, it will seem to have little bearing on their decisions to prosecute areas of inquiry.

The ecological disaster unfolding in our age creates a moral imperative that such a situation should alter. *Nanogovernance* (a term describing the coherence of nanotechnology and science-based natural law governance theory at the small and big – at individual, community, national and global levels) is vital for the sustainability of healthy human societies and the resilience of ecosystems on Earth. *Nanogovernance* promotes use of nanotechnology at local and global levels to alleviate many of the major problems associated with human overpopulation and destruction of ecosystems. Developing *nanogovernance* particularly through a global NES project matters – without it humanity and our environment may not long survive, or have the capacity to flourish.

3. Obstacles to nanotechnology for environmental sustainability

In modern physics, symmetry has proved a fruitful guide to predicting new forms of
matter and formulating new, more comprehensive laws.
– Frank Wilczek, *The Lightness of Being*

A contact with reality, light and intense like the touch of a loved hand ...
the greatest creation of mankind – the dream of mankind ...
Summoned to carry it, aloned to assay it, chosen to suffer it and free to deny it,
I saw for one moment the sail in the sunstorm,
Far off on a wave-crest, alone, bearing from land.
– Dag Hammarskjold, *Markings.*

3.1 A GLOBAL PLUTOCRACY OF SUPRANATIONAL CORPORATIONS

When I was a medical student at Newcastle in NSW I became involved in
civil disobedience protests designed to stop logging by the Forestry Commission of old growth forest at Chaelundi. The fate of these beautiful
hardwoods was to be woodchipped, sold to a supranational corporation
and shipped overseas to be made into paper products by low-cost labour,
then exported back for sale to consumers in supermarkets.

One of our subversive activities was to drive to Misty Creek, walk at
night to the 'feral camp' deep in the woods, then sneak into a logging coupe,
past the circling, head-lit police cars. There, we would erect a wooden
tripod over a logging track to which a protestor then was chained. Next
morning, in order not to hurt the protestor during his removal, a 'cherry
picker' extension crane had to be brought in by the police from some
distance away. Such activities slowed up the logging process until lawyers
were able to find an endangered creature (the HRH mouse – affectionately
known as the His Royal Highness mouse) that finally justified a judge
declaring wilderness protection for the forest.

The willingness to protect our natural environment locally and globally,
as exemplified in such protest actions, appears to have replaced the civil
rights and anti-war movements as a dominant focus of youthful idealism.

I've argued earlier that this, from a more eternal perspective, can be viewed as humans striving to achieve coherence with universal patterns of symmetry and harmony exemplified in nature. If it does, why is this not a generally accepted understanding? Part of the answer may be that such rejection is also part of nature – an inevitable polarity that can be modified, but has to be accepted.

Jean-Marie Gustave Le Clézio's *Ourania* (2005) is a literary example of humans yearning for such harmony, but being opposed by other humans from achieving it. It is set in a remote valley in Mexico, where the main character finds a colony of seekers who have regained the peace of the golden age and laid aside civilisation's ruined customs, including its languages and ways of greed, conflict and war. The same author's *Raga: approche du continent invisible* (2006) documents an idyllic mode of life on the islands of the Indian Ocean, fast disappearing as a consequence of corporate globalisation.

How is it, with all our concern about democracy and human rights, our growing interest in the welfare of other creatures and the sustainability of our shared natural environment, that we seem so powerless to make this the dominant concern of our global governance structures? The answer might be that we have allowed an imbalance to occur – by withdrawing critical governance controls we have let much of our world fall under the control of artificial persons who, motivated chiefly by profit, are prepared to risk destroying it.

Nanotechnology already is big business for supranational corporations. Global government spending on nanotechnology is now about 10 trillion US dollars. That doesn't seem much in relation to the carefully scripted bail-outs of the greed-ridden financial sector we've seen in recent times, the size of the US national budget deficit, global military expenditure, corruption, third world debt or the cost of environmental degradation, war, famine and poverty. The main players in the revolution that is soon to make nanotechnology a ubiquitous part of global human civilisation are supranational corporations investing in the US, Japan, Germany, Russia, India, China and the European Union.

As a result, areas related to nanotechnology regulation that might otherwise be chiefly focused on the public or environmental interest appear unduly influenced by private, corporate concerns. The *International Organization for Standardization*, for example, has been working on definitions of nanotechnology that are important for both safe, efficient manufacturing processes and regulatory assessment of nanomaterials. It has published ISO/TC229-Nanotechnologies, but the document is not freely available to the public, despite the large amount of taxpayer funds that went into its creation. TC229 established a sub-group Working Group 3 on

Health, Safety and the Environment. Those discussions are not being conducted transparently or accountably – no minutes are posted in the public domain and the representatives are heavily weighted towards those in private industry.

From the informal information that is available, corporate representatives in these discussions seem to be striving to develop definitions of nanotechnology that will facilitate rapid regulatory approvals or ensure that cartels of only the largest corporations are allowed to dominate the field. Suspicions that 'business-as-usual' rather than public safety and environmental protection is the dominant ISO concern are heightened by the secrecy of deliberations. Such secretive regulatory processes will form an obstacle to public support for a global NES project.

The capacity for corporate stakeholders to dominate these regulatory discussions on nanotechnology for private purposes is exacerbated by inadequate expenditure of public funds in this area. Each US science and public health agency, for instance, has to make resources available for participation in the maintenance of international nanotechnology regulatory standards from its base budget. This may be a factor in US occupational health exposure limits failing to mesh with those produced by international organisations such as the ISO and OECD for risk management of nanotechnology.

It has been argued in the preceding chapters that primary social virtues are sustained by the collective actions of human beings in overcoming obstacles to act upon principles capable of universal application. The motive for this is conscience, and the ever-widening circle of sympathy it acknowledges has been a major factor in law reform – driving legislative change that allows societies to feel they are characterised by virtues such as equality, justice and (we have argued) environmental sustainability.

A major obstacle to the creation of a global NES project is likely to be that some of the most influential people controlling the research and development of nanotechnology globally do not have a clearly discernable conscience, or the capacity to develop virtues. This is not because they are evil, but because they are artificial persons, supranational corporations, legally created to prioritise shareholder profit above all else. In the 1970s there were 7,000 such artificial persons in existence; within three decades this figure had risen to 40,000 and they collectively controlled one-third of the world's wealth and economic output.

The legal requirement that supranational corporations prioritise maximisation of shareholder profit presumptively isolates them from allegiance to any putative domestic or global social contract. Such artificial persons, though they may be able to sue for defamation, have a conscience only indirectly through that of their directors, shareholders or regulators. They are not allowed to marry, raise children or vote in democratic elections, but they are permitted to influence (through donations, lobbying and media control) legislation such as that on same-sex marriage, the education and health of children, as well as political candidates and their policies. Hitherto we have lacked the wisdom to legally require corporations to have a conscience (or to act towards a nominated public good as if they do) even though such a position appears to cohere with fundamental understandings of social contract, rule of law and science-based natural law theory.

A global project focused on nanotechnology-based environmental sustainability, as Noam Chomsky would say with his characteristic moral force, will be up against influential corporate zealots and proselytisers of free market ideologies who wish to stash humanity into little boxes of consumer contentment manufactured from mass media advertising and sensationalised political distraction, shopping malls and indebtedness, while denying those citizens the information and access to political power needed to participate effectively in a democracy. Media moguls such as Rupert Murdoch and Silvio Berlusconi, for example, have dominated news distribution and used their media empires to provide the public with (often deliberately lurid) stories designed to portray the opponents of corporate globalisation or the supporters of environmental sustainability (such as Green political parties) in a critical light. In return for promoting the election campaigns of pro-corporate and anti-environmental sustainability politicians they appear to receive special consideration in acquisition of new media outlets, in taxation and ability to review impending policy changes.

The worldwide industrial and transportation reliance on oil (an old photosynthesis fuel) highlights many problems with corporate globalisation. As Ralston Saul points out, the dominance of oil in the mechanism of corporate globalisation illustrates not only that market price in such a system is tenuously related to cost, but also that such strategic commodities cannot be fairly priced by market competition and their relative scarcity promotes war. Nevertheless, despite the problematic impact on societal virtues of our need to acquire vast volumes of it, oil remains an ideal corporate product for global profit. People willingly pay large amounts to consume it on the way to buying more and readily become reliant on it to perform basic tasks and obtain enjoyment from life.

There is no doubt that investment by supranational corporations in developing nations facilitates valuable transfers of technology, worker education and employment at higher than local pay standards, with concomitant increases in productivity. Yet, the dominance in global governance systems of supranational corporations legally required to prioritise shareholder profit is significantly distorting the rationality of critical debates related to public health and environmental sustainability. If, for example, a policy-maker now begins to think 'we should pass legislation for universal health care' or 'we should enact a statute that requires inventions be organised towards environmental sustainability', he or she will be met with objections from civil servants or corporate lobbyists that this would contravene obligations under bilateral or international trade or investment agreements.

In many nations supranational corporations orchestrate political candidates and their electoral campaigns so as to maximise capacity to acquire resources and cheap labour and minimise government regulation and taxation. In some poor countries, for example, ambassadors representing nations whose clothing or manufacturing companies rely on cheap labour for global sales have lobbied against an increase in the minimal wage. Politicians become expert performers in enunciating public concerns without being able to effectively deal with them if in doing so they confront the interests of what is in effect becoming a global plutocracy of supranational corporations.

Supranational corporations are a force opposing a global social contract such as may support a global NES project chiefly because they are legally required to prioritise the narcissistic interests of the organisation ahead of the normal range of human concerns expressed in such a pact, including the desire to develop virtue. Global social domination by corporations makes individual citizens secondary and even vicarious participants in democracy.

There are high stakes for supranational corporations should the nanotechnology revolution begin to head towards governance systems that favour environmental sustainability. Many global corporate entities may be concerned that the nanotechnology they are rolling out, ultimately, will take financial and political power from their hands and return it to individuals and communities. Just consider, for example, the capacity of nanotechnology to make nanofactories that produce clothing, buildings, household appliances and transportation devices in a domestic setting. Families and individuals would pay for the equipment and the software then make for themselves whatever they want out of a much wider range of nano-sized raw materials. The profitable but energy wasteful paradigm of corporate globalisation involving rail, road, air and sea transport of raw materials and products would gradually be replaced once cheap domestic nanofactories become a dominant global technology at the service of individual humans and communities.

Unfortunately, at this time of crisis for global environmental sustainability our governance systems seem willing to give corporations greater autonomy from legal obligations to support our primary social virtues and principles. Global governance systems, instead of focusing on environmental sustainability, seem characterised by a transformation from government funding and control in areas such as health care, water, sanitation, power supply, agriculture, banking and communications, to one of lender of last resort and propper-up of failed private sector involvement. From the world's principal creditor in the post Second World War era, the US for instance has become the world's greatest debtor under monetarist policies designed to re-route capital flows back to the US market.

It is no coincidence that the dream of corporate-driven extra-terrestrial terraformation (on the moon or Mars, for example) has been seriously proposed as a policy option for human sustainability at a moment in history when capitalist modes of production are literally tearing the sustainable skin off (and heart out from) the Earth. Nor is it incidental that the life sciences are promising to invent new forms of life at a time of accelerating bioextinction rates. Studies such as the *Limits to Growth* report confirm how the process of corporate profit-driven economic growth is undermining the capacity of the biosphere to support the human species.

Capital constraints are likely to be perceived as a major governance issue for a global NES project in the developing world. A lack of awareness and understanding of NES technologies, for example owing to poor design or limited features, can deter necessary investment. The financial benefits of a global NES project may not accrue to those required or encouraged by policy to adopt the new technology. For example, landlords may have no incentive to invest in buildings made energy-efficient by NES devices when the benefits go to tenants, and manufacturers will not adopt such technologies unless consumer surveys reveal a willingness to pay for them. The lack of capabilities and capacity in many developing countries to design and implement the required NES-coherent regulations, financing mechanisms and measures is likely to be a further obstacle. Even given sufficient capital, many potential purchasers and investors may not be able to capture the full range of NES device savings, because they lack the necessary implementation capabilities.

One important reason, however, why it is important to overcome such corporate obstacles to nanotechnology-based environmental sustainability is that, without these changes in global governance arrangements, life itself and the destiny and dreams of the human species may become annexed within the unsustainable confines of corporate process.

3.2 INTELLECTUAL MONOPOLY PRIVILEGES AS INALIENABLE RIGHTS

One of my first experiences fighting the capability of supranational companies to undermine a social contract involved arguing before a Senate committee against inclusion of provisions altering Australia's Pharmaceutical Benefits Scheme (PBS) in the so-called bilateral free trade agreement between Australia and the United States (AUSFTA). This was at the time when President George Bush had permitted the US Trade Representative (USTR) to become virtually an agent of US pharmaceutical corporations in such trade agreements (it still is). In reality the AUSFTA could more appropriately have been termed a 'unilateral' and 'restrictive' trade agreement. In the former instance because the obligations were mostly on one side and in the latter case because Australian exporters had to use US agencies that bogged them down in expense and red tape, key Australian produce (such as meat and sugar) was denied efficient US access and Australian markets suffered reduced competition through the imposition of longer and more restrictive monopolistic patent protections particularly for pharmaceuticals.

As a neophyte academic I received a lot of advice that the way to get research grants was not to attack your own government. Yet my government was allowing foreign corporations to alter the PBS through a trade agreement. My conscience couldn't accept this. The PBS was one of the few pieces of Australian public health policy with unquestioned democratic legitimacy, having arisen as the result of a constitutional plebiscite. It had matured into a system that used scientific evidence to value the social innovation of a new medicine and thence to initiate a price negotiation between the manufacturer and a government with monopsony buying power. Now the USTR was arguing that a key component of the PBS had to be changed so that pharmaceutical 'innovation' was only valued through the operation of market forces. Further, the USTR maintained, a whole raft of changes to patent law had to be made under Chapter 17 of the same trade agreement, despite the fact that these were pro-monopolistic provisions that actually inhibited competition and market access. Contrary to basic social contract principles, the Australian people had no direct capacity to witness or respond to such USTR claims. The trade deal was being negotiated in secret by bureaucrats.

This is part of what I stated to the Senate committee:

> If we go the other way – if we say … 'No, this is an unbalanced agreement; it is all to the reward of pharmaceutical company innovation; it does not reflect the unique egalitarian principles of Australian society' – I think we will be doing a

remarkable service to the people of developing countries. We will be sending a signal of hope to all those people who are being oppressed by exorbitant pharmaceutical prices ...

The civil society challenge to the AUSFTA, its alteration of the PBS and inclusion of pro-monopolistic intellectual monopoly privileges (IMPs), of course, failed. A politician on the Senate committee investigating the issue told me after one parliamentary hearing: "Son, we never say 'no' to a trade deal."

The point being made here is that another primary obstacle to a successful global NES project is likely to be excessive intellectual property laws, particularly patents, applying globally to its core components as a result of pressure exerted through the threat of trade sanctions under bilateral, regional and World Trade Organization (WTO) multilateral trade agreements.

The governance trade-off involved in a society granting a patent involves the encouragement of rapid public dispersal of innovative knowledge and products in return for a legally protected period of market monopoly. In terms of fitting within a social contract, patents should be regarded as primarily a temporary monopoly privilege (to which judicially enforceable legal rights secondarily attach) designed to compensate an innovator (not merely a patent purchaser) for rapid distribution of valuable information.

The rules and laws of intellectual property have generally failed to achieve optimal balance between these two socially valuable ends: reward of innovation and diffusion of knowledge to the public. The philosophy underpinning such rules was once described as emerging from the normative traditions of natural law theory, or utilitarianism. Such claims are rarely made with any frequency or authority today.

In the late 1960s Nordhaus can be viewed as scoping the place of patents in a social contract by calculating the point at which optimal duration of patent protection balances incentives for innovation against the social losses of monopoly exploitation. He showed how optimal patent life over a product is longer if price elasticity of demand is lower and its social benefit is reduced relative to research and development costs.

Research subsequent to Nordhaus attempted to demonstrate that optimal patent duration should be longer for economic reasons where enforcement is costly or incomplete. Likewise, the case was made that the economic incentive of patent life should be shorter where competitors wasted resources with 'window dressing' inventions merely to improve market share. The traditional Nordhaus model was also contentiously modified to include what is referred to as 'cumulative' or 'incremental' innovation.

Some argued, prophetically given later globalisation developments, that if such reasoning was accepted, the patent monopoly would become a form of rent pursued by competing investors until much relevant and anticipated social benefit had been dissipated through duplication.

Because of its influence on world patent law the US situation is particularly instructive. In the 1980s the decision of the US Supreme Court in *Dawson Chemical Company* v. *Rohm and Haas* influentially overruled prior decisions where judges deprecated patents as disguised, socially disadvantageous monopoly rights. The court now declared that 'the policy of free competition runs deep in our law [...] but [that] of stimulating invention [...] underlies the entire patent system [and] runs no less deep'. In *Haas*, reward of 'innovation' through state grant of protectionist monopoly rights achieved the status of 'equal footing' with the previously antagonistic concept of 'free market competition'. In 1982, the US Court of Appeals for the Federal Circuit (CAFC) was created to centralise patents, tariff and custom, technology transfer, trademarks, government contracts and labour disputes within one specialist jurisdiction. Critics feared the new court would be prone to isolation from broader normative systems and to influence by corporate interest groups. Yet these were probably two of the main reasons for its creation.

The CAFC has since, as expected, developed an extremely pro-patent jurisprudence rarely mentioning the word 'monopoly', readily granting large-scale compensatory damages and permanent injunctions, whilst consistently upholding the interests of alleged innovators over purported copiers or generic suppliers. Industry lobbying succeeded in promoting a federal economic policy positing level of output, rather than amount of competition, as the dominant regulatory end point. This allowed pharmaceutical companies in particular to promote high levels of market concentration as efficiencies, rather than price-distorting monopolies and cartels that conflicted with ethical and legal obligations to promote competitively low prices in the public interest.

Intellectual property rights (as they were increasingly called) provided such corporations with a strategy to protect investments and increase revenue, if need be by excluding competition from the market. This was quite different from earlier conceptions, which stressed the role of patents in social diffusion of knowledge. US pharmaceutical companies successfully prosecuted the policy argument that any form of governmental restriction on their prices in foreign countries was an unjustified interference in the marketplace, rather than an ethically and legally legitimate public health restraint on a protectionist market distortion. This was largely an ideological debate with very little objective evidence of public health or environmental impact being adduced either for or against the proposition.

Unfortunately, corporations have often seen the acquisition of a patent by purchase from an inventor (or that person's start-up company) as giving the equivalent of an inalienable natural right like freedom of expression, association or the right not to be arbitrarily deprived of life by the government of one's society. Some executives of pharmaceutical companies argued, for example, that to make it worthwhile to develop new but unprofitable antibiotics or medicines for children, their organisations should be granted perpetual patent rights.

With particular relevance to their impact on a global NES project, laws such as the Bayh-Dole legislation in the US (allowing universities to profit from patents and to grant those patents to private entities) have been accused of suppressing important research and causing the withdrawal of patented products from the market, not because those products had no public or environmental benefit, but simply because that suited the business plans of corporations. One notorious and pertinent example involved the purchase of patents over electric cars in the US to allow their destruction and reduce their competition to petrol-powered automobiles. The extent to which patents increase this disjunction between the needs of public health and the motivations of the research community is one of the biggest problems that must be faced in the establishment of a global NES project.

Intellectual property laws also loom as a major potential impediment to speedier transfer from developed to developing nationals of the technology outputs of a global NES project. This dichotomy between the urgent need for the transfer of technology and the right of intellectual property holders is addressed in the United Nations *International Covenant on Civil and Political Rights* (ICESCR). While Article 15(1)(b) recognises the right of *everyone* to 'enjoy the benefits of scientific progress and its applications', Article 15(1)(c) protects the moral and material interests of authors of any scientific, literary or artistic products.

Such a provision confirms that under public international law governance of intellectual property involves striking a *balance* between private and public interests. In its *Statement on Intellectual Property and Human Rights*, the United Nations Committee monitoring ICESCR compliance by nations likewise emphasises that 'the realms of trade, finance and investment are in no way exempt from human rights principles'. That Committee further points out: 'the end which intellectual property protection should serve is the objective of human well-being, to which international human rights instruments give legal expression'. This is fully in accord with the social contract model.

On 9 July 1982 Barry MacTaggart, then chairman and president of the supranational pharmaceutical company Pfizer International, published an op-ed piece in the *New York Times*. MacTaggart alleged that US knowledge and inventions were being 'stolen' by particular foreign governments through laws encouraging cheap generic medicines. The World Intellectual Property Organization (WIPO) was criticised for 'trying to grab high technology inventions for underdeveloped countries' and for contemplating treaty provisions that would 'confer international legitimacy on the abrogation of patents'. The call was made for trade law to protect patents (on threat of trade sanctions).

The manner by which international trade law may provide an obstacle to a global NES project by supporting a hybrid corporate–natural rights view of intellectual monopoly privileges (such as patents and trademarks) is considered in the next section.

3.3 INTERNATIONAL TRADE AND INVESTMENT LAW CONSTRAINING DEMOCRATIC GOVERNANCE

In the summer of 2009 I stayed in the village of Hermance near Geneva with my family under a Brocher Foundation residential fellowship. Out of the open window of my office, beyond the tall leafy trees, the boats gently rocking on Lake Leman, I could see villages, farms, churches and cloud-capped mountains. Somehow that view stimulated me to think hard about the different ways nanotechnology could assist global health. Those ruminations laid the foundations for this book. They spawned the idea that nanotechnology of a particular kind could be a medicine for the health of the whole planet. One of the more remarkable experiences of living near Geneva is to see how close the CERN basic physics centre and the United Nations buildings are physically to the headquarters of the World Trade Organization (WTO). I found this particularly poignant given how unnecessarily normatively distinct and even contradictory these three international institutions have become.

International trade and investment law (coordinated in particular by the WTO) is the engine room of corporate globalisation. Corporate globalisation is a process by which foreign capital takes advantage of abundant natural resources (particularly timber, oil, coal and minerals) or cheap labour to manufacture products for distribution and profitable sale throughout the world using road, rail, sea and air freight transport, reduced tariffs and mass marketing techniques. If a macroscience NES project is to make a successful contribution to public and environmental benefit, then established thinking suggests it must be rolled out using this 'free trade' and corporate globalisation model.

The norms of international trade and investment law governing corporate globalisation do not sit easily within established social contract, rule of law, or science-based natural law thinking. There are many reasons for this. One is that international trade and investment law is a normative scheme protecting a limited range of corporate-focused interests that does not cover the full range of human societal and environmental concerns. A second is that it represents law at the service of private corporate interests that has never emerged from protracted social contract thinking – its democratic legitimacy rests chiefly on an indirect link to the representatives of nation states who have rarely if ever sought a democratic mandate about its activities. The third is that its governance mechanisms are not transparent or accountable to international 'civil society' or the rule of law. The discussion that follows highlights some important ways in which international trade and investment law may create obstacles to the successful distribution of technology and products from a global NES project.

The WTO, as mentioned, is headquartered in Geneva in a building of suitably brutalist architecture near many of the United Nations human rights organisations with which it normatively has so little in common. The WTO comprises a secretariat and public officials from nation states who have been involved in the creation and supervision of trade and investment agreements. By those agreements states agree to not merely reduce various trade barriers, but also to allow supranational corporations to take control of major national assets (such as intellectual property, hospital and health services, water, agriculture, power generation and manufacturing) in a way that is very hard to undo (owing to the compensation to corporate stakeholders that must be paid by taxpayers).

What has been created in other words is a supranational corporation-controlled legal system that is pushing global governance in directions different from those of democratic-based community and civil society institutions committed to societal virtues such as justice, equity and, increasingly, environmental sustainability that would underpin a global NES project.

One example of a WTO agreement that may create particular problems for the global roll-out of output from a macroscience NES project is the agreement on *Trade-Related Aspects of Intellectual Property Rights* (TRIPS). TRIPS is a pro-patent agreement likely to increase the price paid by governments, communities and citizens for nanotechnology-based products, by requiring increased patent terms and enhanced protection of patent monopolies under threat of trade sanctions. Its norms can be relied upon by corporate lobbyists to restrict the capacity for governments to issue compulsory licences and mass-produce cheaper versions of patented NES products in public health emergencies. The WTO *General Agreement on Trade in Services* (GATS) likewise allows small cliques of government trade officials (many of whom have been appointed from, and/or will subsequently be rewarded with, lucrative private sector employment) to 'liberalise' various health-related service areas (such as hospitals, electricity and water utilities) that otherwise may be amongst the first to receive government subsidies, tax concessions and other incentives to adopt NES project products.

'Liberalise' is a word that draws on liberal ideologies of individual freedom, but in this WTO usage it represents pro-corporate 'spin' to disguise a process that in effect facilitates the ownership of such services by foreign-based private corporations with little local accountability or motivation to reduce costs to citizens. These WTO agreements arose despite considerable evidence against the public benefit of applying pro-privatisation zealous-free-market economic theory to the health and environment sectors. Missing from such sectors, for example, are some crucial components required by these theories: for instance a genuinely competitive market, government capacity to regulate to prevent market failure, or the ability to accurately place a financial value on interests such as good health or a sustainable supporting ecosystem.

Other WTO multilateral agreements that may create obstacles for a global NES project include the *Agreement on Agriculture* (AoA), the *Sanitary and Phyto-Sanitary Agreement* (SPS) and the *Agreement on Technical Barriers to Trade* (TBT). In the period 1970 to 2000, in order to obtain leniency on national debt repayments, many less developed nations were coerced into removing trade barriers via the Structural Adjustment Policies (SAPs) of the International Monetary Fund (IMF) and World Bank. SAPs were a practical manifestation of the so-called 'neoliberal political-economic consensus' that recommended deregulation of financial institutions and government technology regulators, so that free market forces could operate in more lucrative pro-monopolistic conditions.

In practice, neoliberal economic policy and SAPs entailed reductions of government expenditure on health, welfare, education and other public services; privatisation of government enterprises and utilities; reducing government tax revenues; elimination of tariffs and subsidies (in practice for developing nations but not for protected agricultural industries in developed nations); undermining laws for minimum wages, collective bargaining, unfair dismissal and improved employment conditions; opening of capital and currency markets; removing barriers to foreign direct investment; and promotion of private property rights over natural resources and public goods. Such WTO policies, in other words, cut across, oppose and defeat policy initiatives and domestic legislation emerging out of established social contract understandings predicated on foundational social virtues such as justice, equity and environmental sustainability.

The trade-negotiation round of the WTO known as the Doha Round stalled, for instance, because nations of the global south began to call for the implementation of 'fair trade' rather than neoliberal 'free trade' and would not accept demands such as that for a Multilateral Investment Agreement (MIA) allowing supranational corporations to sue governments if policies or legislation (however otherwise coherent with that nation's social contract) impeded investment. The developing nations also opposed such WTO negotiations because they wished to prioritise food security rather than access of their food exports to foreign markets.

WTO agreements do contain some recognition of public health and environmental norms such as those likely to underpin a global NES project. Article 27.2 of TRIPS, for example, provides:

> Members may exclude from patentability inventions, the prevention within their territory of the commercial exploitation of which is necessary to protect ordre public or morality, including to protect human, animal or plant life or health or to avoid serious prejudice to the environment ...

Likewise Article XIV of GATS provides:

> ... nothing in this Agreement shall be construed to prevent the adoption or enforcement by any Member of measures:
>
> (a) necessary to protect public morals or to maintain public order;
>
> See footnote 5
>
> (b) necessary to protect human, animal or plant life or health;
> (c) necessary to secure compliance with laws or regulations which are not inconsistent with the provisions of this Agreement including those relating to:
>   (i) the prevention of deceptive and fraudulent practices ...

Article XXb of the GATT (adopted in 1947 and incorporated into WTO Agreements in 1994) similarly allows an exception to GATT corporate privatisation rules when that is necessary 'to protect human, animal, or plant life or health'. Exceptions along these lines are now found in the *Agreements on Application of Sanitary and Phytosanitary Measures* (SPS agreements) and the *Technical Barriers to Trade* (TBT) agreement.

Some commentators try to extol these limited public interest exceptions to WTO agreements as if they were the normative equivalent of the great

United Nations international human rights conventions. Instead, the reality is that neither WTO agreements or other regional or bilateral trade and investment agreements specifically refer to human rights norms or have dispute settlement panels composed of members with any expertise, inclination or textual mandate to apply them.

To give an example, in 1988 the European Union (EU) imposed a ban on the sale of beef from cattle fed with artificial hormones following the precautionary principle and evidence that this could cause cancer or nerve disorders. The US challenged the decision and a WTO panel of trade lawyers ruled the ban was illegal (against the restricted set of WTO norms they apply) chiefly because it was inconsistent with the SPS agreement and its risk-assessment procedures.

The United States has a long history of using bilateral and regional trade agreements to influence health and environmental policies in other nations to the benefit of its corporations. In 1988, for example, an amendment called 'Special 301' was made to a section of the *Trade Act 1974* (US). This became the principal statutory authority under which the US investigated and, if need be, threatened trade sanctions against foreign countries that maintained acts, policies and practices that its corporations considered violated, or denied their rights or benefits under trade agreements, or (through otherwise being justifiable, reasonable or non-discriminatory) nonetheless burdened or restricted US commerce. The USTR was required under the *Trade Act 1974* (US), to create, in its annual review, a Special 301 Report Priority Watch List. Using this mechanism corporations can petition the USTR to investigate and, ultimately, threaten trade sanctions against what they perceive to be an unjustifiable, unreasonable or discriminatory global NES project-related policy of a foreign country (for example a subsidy for NES products that were competing in the market against existing patented products).

More recently United States-based corporations have been instrumental in inducing the US Trade Representative (USTR) to negotiate a series of regional and bilateral Free Trade Agreements (FTAs) in which provisions are included for increasing intellectual monopoly privileges (IMPs), promoting investor–state dispute settlement mechanisms and pressuring health-technology cost-effectiveness assessment systems in ways not possible in the WTO (where the bargaining power of the US is countered by opposing blocks of developing nations). These mechanisms too could become obstacles to a global NES project that develops, for example, nanomedicines or nanotechnology-based medical devices.

The capacity of US bilateral trade agreements to undermine a global NES project is highlighted by their use to attempt to alter other public-focused regulatory processes to which nanotechnology-related products could apply such as quarantine and blood fractionation. A World Health Organization (WHO) commission and numerous civil society publications have documented the contradictory relationship of such provisions with the *Doha Declaration on TRIPS and Public Health* and their potentially deleterious impacts on public health. Yet such trade deals continue to proliferate – often tying up the exporters of signatory nations in red tape and requiring pro-monopolistic and anti-democratic changes in their domestic legislation.

A notable example of this use of bilateral trade agreements to shape the regulatory architecture applying to new technologies away from what might have been predicted by social contract analysis involved the combined operation of provisions in the *Australia–United States Free Trade Agreement* (AUSFTA) Chapter 17 (on intellectual monopoly privileges) and Annex 2C (on changes to the Australia's Pharmaceutical Benefits Scheme (PBS)). Australia's PBS system involves experts in pharmaco-economics assessing the scientific evidence of the cost-effectiveness of a new patented medicine then making a recommendation as to whether it should receive a government subsidy. The Annex 2C changes established a legislation-lobby group that ultimately produced legislation that substantially reduced science-based cost-effectiveness or reference pricing in the PBS as US pharmaceutical companies had demanded.

The peak US pharmaceutical industry lobby group (PhRMA) has used other bilateral and regional trade agreements to seek governance changes in other nations that ensure its members' intellectual monopoly privileges are not 'undermined by other government pricing and regulatory mechanisms' (such as cost-effectiveness research (CER) and reference pricing systems) which it refers to as 'non-tariff' barriers to 'innovative medicines'.

Supranational companies might use many other aspects of trade and investment agreements to preferentially alter domestic governance arrangements concerning outputs from a global NES project. Apart from specific provisions increasing intellectual monopoly privileges, they may facilitate revolving door appointments (between private interest lobby groups and the USTR trade offices) and requirements to use complex US-preferential procurement agreements, or only use US exporters.

'Investor–state' dispute settlement provisions represent another tactic of particular concern as a potential obstacle to a global NES project. Basically they allow supranational corporations to sue nations before small panels of commercial arbitration lawyers with little understanding of or desire to apply international public law and a vested interest in perpetuating a system where disputes can only be initiated by those corporations. The grounds for the suit usually involve governance structures including legislation (even when in the public health and environmental interest based on good scientific evidence) if the commercial interests of those corporations allegedly are thereby impeded.

Investor–state provisions came to prominence in the 1994 *North American Free Trade Agreement* (NAFTA) between the United States (US), Canada and Mexico. In the 1990s civil society prevented the creation of a supranational investment protection agreement (the *Multilateral Agreement on Investment* or MIA) that would have allowed the global implementation of such provisions, but they have nonetheless proliferated in a series of bilateral and regional arrangements. Nonetheless, investor–state clauses have now become a controversial part of bilateral investment treaties and over 300 investor–state dispute settlement cases have been decided (none ever against the US).

The lawyers officiating on such arbitral proceedings, as mentioned, by training view such investment agreements as private contracts, are paid by the parties and do not necessarily take account of domestic public health and environment protections – creating a pro-investor jurisprudence. It should be of concern to those supporting a global NES project that investor–state challenges have occurred in relation to a broad spectrum of public health and the environment legislation and policies. Supranational corporations could use this mechanism to claim compensation where a global NES project was subsidised by a government on the basis that its products were more environmentally friendly, or safe from a public health point of view. For example statutes on water protection, waste disposal and waste treatment as well as universal health care or access to affordable medicines have been challenged by supranational corporations under investor–state mechanisms.

As a further illustration, in the negotiations for a *Trans Pacific Partnership Agreement* (TPPA) the supranational Philip Morris tobacco company argued for the inclusion of an investor–state provision on the basis that it would assist it to prevent Australia's regulatory moves towards plain packaging of cigarettes. The same company initiated a damages claim against this Australian legislation using an investor–state provision in a Hong Kong–Australian investment treaty that had no exceptions for public health legislation. When the Australian government dug in and passed the legislation, the USTR negotiators to the TPPA argued for a 'carve out' of tobacco to ensure the investor–state dispute settlement provision was included that would allow supranational corporations to hold the signatories as policy hostage in a variety of areas related to public health and environment protection.

This suggests that should a global NES project come up with a product that replaces those upon which supranational corporations have substantial investments (in say old photosynthesis fuels or electricity distribution networks), then those corporations may well resort to investor–state mechanisms to protect their profits and inhibit the roll-out.

Another tactic by which international trade agreements may create obstacles for a global NES project is the non-violation nullification of benefits (NVNB) provision. Highly secretive and poorly understood, NVNB clauses further heighten the suspicion that what has been created in the WTO and bilateral dispute resolution process is not law emerging from a social contract, but law at the service of large-scale private corporate interests.

NVNB claims are directly referred to in Article 26 of the WTO *Dispute Settlement Understanding* (DSU), in the GATT, Article XXIII of GATS and Article 64 of TRIPS. Under such NVNB provisions, the full range of dispute resolution mechanisms may be invoked whether or not a breach of any specific provision in a trade agreement is alleged or substantiated. The precondition is that a 'reasonably expected' 'benefit' accruing under the relevant trade agreement has been 'nullified or impaired' by a 'measure' applied by a WTO Member. NVNB provisions thus cut across two of the foundational social virtues underpinning the rule of law under domestic or global social contracts – certainty and predictability.

Both the United States and European Economic Community have argued before a GATT 1994 panel that recourse to NVNB claims should remain 'exceptional' otherwise 'the trading world would be plunged into a state of precariousness and uncertainty'. Contemporary controversy over NVNB claims and proceedings arises in large part from their potential to allow a WTO Member to tactically exploit a trade agreement's textual 'constructive ambiguities' and threaten a dispute if a wide and largely undefined range of domestic regulatory components is not altered, or compensation organised. Use of NVNB provisions may facilitate a WTO dispute settlement process involving deliberate diplomatic 'gaming' of trade 'rules', from what had otherwise been viewed as finely balanced textual truces (such as between the advertising and scientific definitions of 'innovation' in Annex 2C of the AUSFTA), where uncertainty is deliberate and inherent and dispute panel interpretation more an act of ongoing negotiation, than judicial analysis.

Article 3.2 of the DSU requires panels to clarify existing provisions of agreements in accordance with customary rules of interpretation under public international law. This leads to consideration of how the NVNB principle interacts with Article 26 of the *Vienna Convention on the Law of Treaties*, incorporating the principle of *pacta sunt servanda*: 'Every treaty in force is binding upon the parties to it and must be performed by them in good faith.' NVNB claims appear to undermine this fundamental principle of international law (when considered as a global social contract) by subsequent reinterpretations based on the 'spirit' of the agreement. At the WTO meeting in Hong Kong in December 2005 the United States delegation pushed hard behind the scenes for trade concessions in return for its acquiescence to the moratorium on the use of NVNB provisions under TRIPS. The resultant Ministerial Declaration left the position of NVNB claims under TRIPS extremely uncertain.

NVNB claims, in other words, are another tactic whereby supranational corporations whose interests are allegedly adversely impacted may attempt to inhibit governments from passing legislative subsidies or other measures supporting a global NES project even where that is in no direct violation of a WTO or other trade agreement.

3.4 RESTRICTED DEMOCRATISATION OF GLOBAL NES POLICY

My friend Georgia Miller works for the non-governmental organisation Friends of the Earth (FOE) and has been in charge of various FOE projects in Australia dealing with nanotechnology. Georgia is a vegetarian and a person of strong ethical values. She is convinced there should be a moratorium of all nanotechnology until it is proven safe. At nanotechnology regulation conferences and workshops we often debate this as I promote the idea that nanotechnology can benefit human health and the environment. Georgia is sceptical of such claims. I once asked her why, given that critical stance, she was invited to be on so many government-supported committees and workshops about nanotechnology. She replied that one reason (apart from her expertise) was that her presence, by representing FOE, allowed industry and government to represent that their regulatory and marketing decisions were collective, transparent, accountable and legitimate.

If a global NES project is to be successful it will need the support of international civil society, for instance through the aegis of NGOs (such as Friends of the Earth) that have previously been much more focused on the safety and toxicological problems associated with nanotechnology. Some food and cosmetic companies espouse the view that nanotechnology has already developed an adverse public 'reputation' and have removed all references to nanotechnologies (even if present in their products) from their websites. In Australia some sunscreen companies have inscribed 'not-nano' on their packaging, a move that has been criticised by regulatory officials after complaints by other nano-using cosmetic corporations who felt their investments were thereby being impeded.

Aware that the public has suspicions about the safety of nanotechnology, but not wishing to impede the industrial development associated with it, governments have set up consultative bodies and processes. These are ostensibly designed to obtain input from the community and to reassure the public that nanotechnology is safe. No doubt a global NES project will have to run the gauntlet of such 'consumer-focused' committees and programs in many nations.

Measures that would actually have promoted both these aims in a practical sense have not been tried. These might have included pilot programs allowing concerned citizens to report online (or even arrange for samples to be collected according to standardised protocols) examples of nanotechnology in the marketplace that appeared to raise safety concerns. Examples of marketed nanoproducts that could have been so examined include the use of carbon nanotubes in paving tiles (cut by electric saw for placement), use of nanosilver in aerosol sprays as an antiviral agent or in clothing and washing machines that might wash it into sewerage systems and waterways. They might also have included capacity for civil society to have input into government subsidies and tax concessions designed to encourage research and development of nanotechnology focused on critical public health and environmental problems.

The near ubiquitous use of the term 'consumer' in government-sponsored public consultation documents about nanotechnology seems to dismiss some of the most powerful connections that citizens in a functioning democracy should have to that society's social contract, with all its associated rights, duties and responsibilities. 'Citizen' is a word that coheres much better with social contract theory, the rule of law and the science-based natural law understanding of how a person should participate in their local governance arrangements.

Yet, this is not the present model underpinning the global nanotechnology corporate and government regulatory endeavour. To take one controversial example, labeling of products as containing nanoparticles is opposed by supranational corporations as unnecessarily playing upon public lack of knowledge or fear. But labeling of nanoparticles, even if it does occur, is chiefly held up as enhancing 'consumer' choice – as if such choice in 'buying products' is somehow the apotheosis of democratic achievement under a social contract.

Of course regulators should support initiatives such as creating and maintaining a list of marketed products containing nanomaterials and supervising appropriate packaging labeling of associated risks. Surely, however, this is only the start of proper democratic engagement in the global roll-out of nanotechnology. A successful global NES project would have to receive much wider and deeper support and input from citizens.

Although this may seem an unusual point to make, I feel that a major obstacle to democratic involvement in a global NES project will be the power of elected legislatures. This is because the legislative process in many economically developing and developed nations appears to have been captured and manipulated by self-interested and/or corrupt private interest groups. Lobbyists for supranational corporations and the political actors and trade negotiators beholden to them, for example, take advantage of executive fiat to rule with only intermittent and vague mechanisms of accountability, particularly where constitutional texts and a corrupt judiciary deny citizens the capacity to challenge the state in human rights actions.

A second problem for democratic involvement in a global NES project will be that the 'positivist' justification for the necessary legislation (including its much vaunted support for the social virtues of consistency and predictability) is founded on what is in reality a fiction. This is that the majority of citizens in a nation, or on Earth, by their free will have consented to the national constitution or international convention (the domestic or global social contract respectively) that creates a foundational social 'rule of recognition'. Mechanisms now exist that would readily facilitate all citizens voting on core components of such social contracts or their legislative or treaty outputs. Instead, however, the will of such citizens is considered by the dominant positivist jurisprudential theory to be expressed vicariously though the activities of (too readily lobbied or corrupted) elected politicians or the government officials they appoint.

Such anti-democratic governance mechanisms may impede a global NES project if they reduce its capacity to appeal to public interest on broadened, deepened and variable spatial and temporal scales, including threshold and absolute ecological limits to human activities. They may also hold back a global NES project from cohering with emerging patterns of symmetry in human–environmental relations including prioritising mechanisms for addressing risk and uncertainty and adopting a governance framework that respects non-traditional moral entities such as non-human species, ecosystems and future generations. Likewise traditional governance structures based on vicarious citizen representation may hinder the ability of a global NES project to address systemic obstacles embedded in patterns of production, consumption, settlement and governance, specify methods for fairly assessing the economic value of its output in terms of community cost-effectiveness, have clearly defined confidentiality and intellectual property arrangements or promote and fund research into its ethical, legal and social implications.

The fact that a group of citizens living on the same land mass and subject to similar political or cultural backgrounds once voted on constitutional arrangements (something, by the way, that did not happen directly in relation to the United Nations Charter) provides a contentious link to legal obligation (except for judges rigidly adhering to extreme 'originalist' views of legal positivism) in the present generation and for something as important as a global NES project. To apply an analogy, the fact that the head of a family several hundred years ago decided that all subsequent family members should follow particular rules of conduct on most conceptions of relations prioritising free will and liberty does not invest such pronouncements with perpetual legitimacy. An exception might be where ultimate religious authority is associated with such a sire and those who follow remain acquiescent to the associated teachings.

In conclusion, it will be a major obstacle for a global NES project if domestically and internationally our political leaders fail rapidly to put in place institutions and processes that facilitate global society rapidly and efficiently championing environmental sustainability as a foundational virtue.

3.5   PUBLIC AND POLITICAL DISTRUST OF A GLOBAL NES PROJECT

Bitter experience has taught many citizens and community groups to be sceptical of government or supranational corporate claims that products are safe, that water, soil or air is not polluted, that the financial sector is properly regulated, and that key elements of social services (such as hospitals, schools or the police) will not be sacrificed in budgetary discussions in favour of profits by real estate developers or corporate entities. They are likewise suspicious of assertions that water and power utilities, or health care systems will work more efficiently when handed over to a private sector that each year will seek to increase the cost to taxpayers by a few percentage points more than the inflation rate.

This scepticism and suspicion of the market-state governance system in developed nations has disastrously come to the fore in the contemporary debate over when and how to take policy action to mitigate the impacts of anthropogenic climate change. Attempts by the corporate sector making large profits from old photosynthesis fuels (for instance oil, coal and natural gas) to delay carbon pricing or taxation by discrediting the scientific experts on the United Nations commission tasked with investigating the question will probably go down in human history as a low point for rationality in the face of a collective crisis.

If scientific research begins to accumulate that a particular global NES project offers the best chance for our species and planet, it is nonetheless quite likely (given the scenario played out in relation to climate change) that NES 'deniers' will protest against the 'vast' amounts of money being dedicated to the project when governments should be trying to reduce debt, or spending on defence as well as other components of social infrastructure. Those supporting a global NES project will have to defeat or conciliate socially powerful political oligarchies and, for instance, old photosynthesis fuel industry lobby groups that will be trying to convince 'consumers' to trust it is in the public interest to oppose a global NES project that offers cheaper forms of renewable energy.

Policy-makers may not prioritise the funding or institutional support for such an NES project because of competing goals related to personal or party short-term electoral success. There may be claims they preferentially have to address, for example job losses in old photosynthesis fuel-based industries, loss of strategic control of already heavily invested-in power infrastructure, and inhibition of foreign investment in established non-nanotechnology-based industries.

Competing academic research sectors could also become resentful of funds flowing in this new direction and criticise the NES project for lack of clarity in research questions and policy objectives as well as lack of consensus on the supportive research evidence. Such competing scientific voices will provide metaphoric ammunition for those opposed to a global NES project of any description, if only simply because they dislike the idea of new technology flowing equitably throughout the world to help solve its major problems, given that those very problems (for instance war, ill health) are often a lucrative indirect source of corporate income and research funding.

The public will distrust a global NES project if the mechanisms set up for their participation for instance are viewed as tokenistic (pretence of informing the public about major NES decisions), manipulative (public informed of major NES decisions only when convenient), passive (public informed about major NES project decisions after they are taken) or scripted (by government or corporate public relations consultants associated with the global NES project).

Likewise, a global NES project may not engage the public unless those corporations involved are encouraged or required by law to have environmental sustainability duties. An example of this type of obstacle would be if corporate involvement in the selected global NES project (according to the specified governance requirements) could not be included as a highly ranked contributor to governance measures such as the Dow Jones Sustainability Index (DJSI). Another example could involve tax concessions or subsidies for investment in a global NES project.

In broader terms, a substantial obstacle to a global NES project could be the lack of robust domestic and international mechanisms for developing long-term visions of the sustainability of human society and its supporting ecosystems on Earth. This is compounded by the lack of serious policy debate conjointly at international (United Nations), governmental and community levels of global governance about how each human family should live, the type of ecosystem to surround them, the number of people and the number of other species on Earth. This debate needs to be taking place at the highest levels of global government as well as in local communities. Our inability to present a consensual optimistic vision of human sustainability on Earth inhibits our capacity to produce leaders willing to step up and take responsibility for achieving it.

For a global NES project to be established both policy-makers and the public may need, in other words, to be both better educated about the scientific evidence justifying that endeavour, and encouraged to give the public equal input with the corporate sector in the decision-making processes. In this sense a global NES project could itself be the catalyst for much wider positive and optimistic governance changes in global society.

4. Core normative components of a global NES project

All the diversities of the world were brought together,
the blessings of nature were collected,
and its evils extracted and excluded ...
nothing is more common ... than to call our own condition
the condition of life.
– Samuel Johnson, *Rasselas*

[I]n principle it is so easy to simulate a universe in a computer
that there must be very many simulations scattered across the Multiverse.
– John Gribbin, *In Search of the Multiverse*

4.1 NES FIRMLY SUPPORTED BY ETHICS AND LAW

Imagine, as a thought-experiment, a repeat of the experiment of the fully enclosed ecosystem Biosphere 2. This time nanotechnology would provide the framework for the artificial Eden and the urgency for its success would be given by the collapse of Earth's ecology under human population growth with related energy and resource pressures.

Nano-Biosphere 3, as it might be known, would be a prototype for a nanotechnology-based world fundamentally committed to environmental sustainability. It might, for instance, involve communities attempting to live inside suburb-sized huge geodesic carbon-nanotube domes to see if they can use various forms of nanotechnology to create a flourishing life. Nanotechnology would be used in such a community for instance to assist the provision of education, food, water, medicines, security and energy. Imagine it as a type of testing ground for the best potential outcomes from a global NES project.

Nano-Biosphere 3 mini-suburbs, like the original Biosphere 2, might each contain different ecosystems now at risk of permanent degradation or extinction: a rainforest, an ocean coral reef, mangrove wetlands, a savannah grassland, a small farm and well-treed urban habitat with vegetable gardens. Power would be obtained from nanotech-based panels in all engineered structures that efficiently trapped solar energy in chemical bonds by using it to split waste water and create hydrogen (burnt to make fresh water) as well as oxygen; and by then also utilising carbon dioxide to make basic starches. Waste water would be purified using nanotube filters, then re-cycled. Food would be grown inside without chemicals, including fruit, legumes and green vegetables, additional nutrients added and spoilage prevented by nanotechnology. The buildings would be carefully designed using multi-walled carbon nanotubes to withstand stresses and facilitate energy, pressure and temperature regulation as well as solar fuel production. Nanofactories would make consumer products from basic materials and software.

There would be a nanotechnology-based artificial wind system and monitoring of atmospheric CO_2 and O_2, ocean calcium carbonate and pH, composting, and which plants were thriving. Excess of atmospheric CO_2 was the major problem in Biosphere 2 (alongside emotional tussles between inhabiting researchers). CO_2 was higher at night and in winter, outpacing the capacity of even fast-growing plants to perform photosynthesis. Bees, small birds and humans failed to thrive inside Biosphere 2 whilst cockroaches did. In Nano-Biosphere 3 nanotechnology would provide the means for carbon dioxide reduction and conversion to starches and fertilisers.

What would be the crucial feature of a successful Nano-Biosphere 3? Such a project might give us important insights into the type of nanotechnology that should be employed to best promote environmental sustainability. It could become a symbolic reference point for how nanotechnology can help shape a better world. But could it also be designed to test the type of ethical and legal norms needed to make sure nanotechnology functions equitably in this task?

As a conceptual background to this, what, for example, if we viewed the cosmos as a type of sustainability experiment by a much more advanced consciousness? Would that help explain established anomalies in both physical and moral laws? Physicists have already postulated that the entire universe (at least that we are aware of or can presently conceive) is so dependent on a series of remarkable coincidences that it could represent a type of mathematically and geometrically based construction.

It has, for example, been a critical physical law for sustainability of this universe that the excited energy levels in millions of electron volts of carbon nuclei (three helium nuclei combined) are tuned just right to resonate with a beryllium nucleus (two carbon atoms), but not with oxygen nuclei (four helium nuclei combined). A similarly vital law is that it takes 10 billion years for stars to scatter heavy elements and for intelligent life to evolve – which is just the point we are at in relation to our capacity to see the distant extent of the universe. Likewise, no carbon-based biosphere could exist if the physical laws determining energy efficiency of helium production from hydrogen, or the strength of gravity and its proportion to the expansive force of the universe were slightly different from what they are.

Returning then to Biosphere 3, one thing immediately clear from such a thought-experiment is the potential isolation its inhabitants would experience from the grandeur of the universe, from the stars, winds, clouds and seasons, from areas of wilderness abundance. Would a conclusion be that we need such exposure to unenclosed nature to temper our self-assurance, moderate our passion, excite our imaginations or spark our moral sensibilities? Might we learn that our capacity to experience and ponder the vastness of space is critical to that moral development that may be central to some significant anti-entropic (complexity-building) role that, at our virtuous best, we fulfil in the universe?

A conclusion might be that, although nanotechnology can help us to survive and contribute to the sustainability of the ecosystem that surrounds us, it is involvement with a world that is natural in the sense of not being made or controlled by man that looms as a crucial component of our flourishing – something that a global NES project should support rather than supplant.

Further assume, as part of the same thought-experiment, that Nano-Biosphere 3 is governed by a published social contract under which ethical and legal norms attempt to prioritise the social virtue of environmental sustainability and related governance principles, rules and processes. From the arguments put forward in earlier chapters, these governance mechanisms could be presumed to operate in their most fundamental sense as resonances of physical laws. Nano-Biosphere 3, in other words, could be explicitly designed to manifestly promote foundational social virtues like justice, equity and environmental sustainability (viewed as harmonics of an inherent universal symmetry and harmony).

More particularly, the published social contract of Nano-Biosphere 3 might support ethical and legal principles developing and disseminating environmentally sound technology (EST). It would be coherent, for instance, with *Agenda 21* arising from the 1992 United Nations *Conference on Environment and Development* (Earth Summit), which defines ESTs as those that 'protect the environment, are less polluting, use all resources in a more sustainable manner, recycle more of their wastes and products, and handle residual wastes in a more acceptable manner than the technologies for which they were substitutes'. The development of ESTs, as governed by the Nano-Biosphere 3 social contract, would require not only innovations in equipment or hardware, but also in the principles, rules and systems (such as corporate organisational and managerial procedures) needed for the preservation of ecosystems.

The social contract governing Nano-Biosphere 3, reinforcing the foundational social virtues of justice and equity as well as environmental sustainability, would raise public awareness of norms whereby the utilised nanotechnology would augment rather than supplant traditional and indigenous knowledge and practices about sustainable living. Indeed, Nano-Biosphere 3 might highlight how such traditional approaches to sustainability (often marginalised by the economic power of the globalised 'old photosynthesis' fuel-based industry and by an increasingly urban-dense population) are important morally in preventing human alienation from the world's life cycles.

The social contract underpinning Nano-Biosphere 3 could also be designed to mesh with a global social contract promoting environmental sustainability-oriented governance, such as assisting international public law to firm up obligations (particularly in the trade and investment law sectors) concerning 'technology transfer'. It would cohere, for example, with the Intergovernmental Panel on Climate Change (IPCC) report on *Methodological and Technological Issues in Technology Transfer* that defines the transfer of technology fostering environmental sustainability as 'a set of processes covering the flows of know-how, experience and equipment, for mitigating and adapting to climate change amongst different stakeholders such as governments, private sector entities, financial institutions, non-governmental organizations (NGOs) and research/education institutions'.

This would involve developing processes to understand, utilise and replicate NES project technology including adapting it to local conditions and integrating it with indigenous technologies. Such published NES project technology transfer social contract obligations could be government-driven (initiated by government to fulfil specific policy objectives), private sector-driven (transfers between sponsoring commercially oriented private sector entities), or community-driven (transfers involving Nano-Biosphere 3-related community organisations with a high degree of collective decision-making).

A Nano-Biosphere 3 prototype social contract should also cohere with normative statements such as that in Article 19(2) of the United Nations *Convention on Biological Diversity,* which stipulates that access to technology by developing countries shall be on a 'fair and *equitable* basis'. Likewise such a charter could highlight how the moral principle of equity (requiring equality of persons – all deserving equal concern and respect) as formulated in such a document, requires NES project outcomes be proportional to input or contribution, as well as coherent with equity in international law that supports claims by developing countries to a fair share of natural resources or access to markets and new technology. Related norms could include developing community claims to be compensated for conservation of natural resources (such as rainforests, biodiversity and crop diversity) and the development of medicinal knowledge from traditional wisdom about plants, which were involved in the creation of or impacted by NES products.

Responsibility is another recognised ethical and international law principle that could be reflected in prototype Nano-Biosphere 3 governance arrangements. It would resonate, for instance, with Principle 7 of the *Rio Declaration*, which, for example, makes the moral case that in view of the different contributions to global environmental degradation, states have common but differentiated responsibilities. The ethical idea encapsulated here is that the Nano-Biosphere 3 governance model supports an onus on industrialised nations to assist developing countries to restore and preserve the sustainability of ecosystems of the Earth by, inter alia, transferring to them environmentally sound technology (such as those from a global NES project).

Another ethical principle potentially derived from such NES project social contract arrangements might be that such obligations should be greater where citizens have enhanced capability to assist others (with minimal risk or loss and where the resultant benefit is proportionately higher). Such a concept of moral responsibility picks up the second part of Principle 7 of the *Rio Declaration*, which, for example, refers to the 'technologies and financial resources' commanded by the developed states. The *Climate Change Convention*, following the wording of *UNGA Res 44/228*, likewise creates a moral obligation on states with greater current capability to tackle the causes of global environmental problems (especially greater access to technology and resources), to assist other states in the implementation of their international commitments. As Principle 6 of the *Rio Declaration* notes, 'the special situation and needs of developing countries [...] should be given special priority'.

In a similar vein the Nano-Biosphere 3 social contract could resonate with the *Copenhagen Accord* in recognising that 'social and economic development and poverty eradication are the first and overriding priorities of developing countries'. It could recognise that only through a substantial increase in the transfer of financial and technological resources can developing states simultaneously improve their socioeconomic situation and reduce their future negative impact upon the environment. It might even highlight aspects of the pro-corporate WTO TRIPS agreement (such as Article 66(2)) that support transfer of nanotechnology products focused on environmental sustainability to developing nations 'in order to enable them to create a sound and viable technological base'.

The *United Nations Framework Convention on Climate Change* (UNFCCC) could provide another point of reference for these prototype global NES governance arrangements. Article 4 provides, for example, that developed country parties (and non-state organisations) are obliged to take 'all practicable steps to promote, facilitate and finance, as appropriate, the transfer of, or access to, environmentally sound technologies and know-how to' developing countries and to support the development and enhancement of endogenous capacities and technologies. The United Nations *Convention on Biological Diversity* likewise in Article 16 obliges states to undertake to provide and/or facilitate access to and transfer of technologies that are relevant to the conservation and sustainable use of biological diversity including by altering as necessary patents and other intellectual property rights.

The international human right to health could also be highlighted as supporting a norm for nanotechnology-based sustainability of ecosystems owing to the close interdependence between a sustainable ecosystem and the state of human health. Another emphasised source of the social contract norms behind a global NES project might be provisions such as those in the UNESCO *Universal Declaration on Bioethics and Human Rights* (UDBHR). Article 14 of the UDBHR, for example, requires that progress in science and technology should advance, amongst other things, access to quality health care and essential medicines, access to adequate nutrition and water, improvement in living conditions and the environment and reduction of poverty and illiteracy. Article 15 of the UDBHR similarly requires that benefits resulting from any scientific research and its applications should be shared with society as a whole and within the international community, in particular with developing countries. Article 21 of the UDBHR requires transnational health research should be responsive to the needs of host countries, and the importance of research contributing to the alleviation of urgent global health problems should be recognised.

The next section explores the importance, in terms of achieving widespread public acceptance as well as coherence with governance patterns emerging from fundamental physical parameters, of a global NES project's governance arrangements supporting notions of individual and community conscience as well as concepts of public and environmental good.

4.2 NES FOCUSED ON A GLOBAL CONSCIENCE, PUBLIC AND ENVIRONMENTAL GOOD

At the completion of my PhD and at the point of making the transition from full-time intensive care medicine to full-time academic chair (of the *Medical Professionalism and Leadership* theme at the new Australian

National University medical school), I was in London living near Hampstead Heath and studying at the Wellcome Library. A main topic of my research was the governance implications of the Bristol paediatric surgery scandal, which had been exposed by the whistleblower anaesthetist Dr Steve Bolsin. One day, while working at the Wellcome Library, I experienced another of those unusual coincidences that, whilst no doubt statistically explicable, over time I've begun to view as significations that my interests may be resonating with an emerging pattern of symmetry.

In my research at the Wellcome Library I had discovered that key statements of medical professional governance such as the *Geneva Declaration* (the modern restatement of the *Hippocratic Oath*) require doctors to follow conscience in the interests of protecting their patients, even if that means disobeying the law. I'd also learnt that the core document of the international human rights movement, the United Nations *Universal Declaration of Human Rights* (UDHR), also supports the primacy of conscience in the fundamental normative constitution of humanity. Article 1 of the UDHR for instance provides 'all human beings are born free and equal in dignity and rights. They are endowed with reason and conscience [...]' Freedom of conscience is one of the inviolable (cannot be contravened by opposing domestic laws) basic international human rights guaranteed by Article 18 of the UDHR.

In this way I researched at the Wellcome Library how conscience in the normative systems of medical ethics and international human rights interacts with legislative rules. I was contacted one day, whilst so engaged, via email by a rehabilitation physician at the Canberra Hospital (Dr Jerry McLaren). Jerry explained he was working with the support of Dr Steve Bolsin (now professionally ostracised from the UK as a result of his successful whistleblowing and working at Geelong Hospital in Australia) to expose the substandard practices of a prominent neurosurgeon that allegedly were maiming and killing patients in my hometown.

I felt strongly motivated in conscience to support Jerry regardless of the inevitable risks to my nascent relationships with professional colleagues. I'd already confirmed on numerous occasions the coherence and expansion of being felt in allowing conscience to provide the motive force to apply universally applicable principles in the face of obstacles; as well as a corresponding inner contraction (despite any short-term material advantages that might accrue) when the call of conscience was denied. We organised a Grand Rounds presentation to publicise the issue, although this aroused legal threats to hospital management and the disapproval of many senior colleagues.

Afterwards, I wrote some articles on the problem of why the system of medical professionalism should valorise conscience in its constitutive texts, but have no standardised place for one of its most obvious manifestations (whistleblowing on clinical negligence and incompetence) in routine governance pathways. I then began to teach the medical students in the course I was running how to whistleblow strategically and systematically, instead of the process remaining an ad hoc activity by professional pariahs somehow sequestered from the clinical governance pathways in which it should be an important but late option.

Deeds of conscience, displaying commitment to ideals about public and environmental good even in the face of persecution, seem to me to reveal a profound truth. My studies of such courageous actions have led me to support the natural law hypothesis that they show presumptively noble human beings striving to make morally complex circumstances cohere with universal patterns of symmetry fundamentally present in the physical world. That's one of the reasons why I started to support whistleblowers when I became an academic. I always felt these people should enjoy a rebuttable presumption that they were exercising conscience to achieve a public good even at the cost of severe disruption to their personal lives.

This section focuses on the importance of a global NES project resonating with individual and community conscience in its pursuit of public and environmental good. The writings of the philosopher Schopenhauer (like those of Kant) seem to me to be very instructive on the issue of how conscience might be conceptualised within a normative system seeking coherence with modern physics. Schopenhauer held that in a healthy mind only selfish deeds oppress our conscience, not wishes and thoughts; for it is only our deeds that hold us up to the mirror of our will.

Schopenhauer claimed that the good conscience we experience as an enlargement after every disinterested deed arises from the fact that such an act verifies the knowledge that our true self exists not only in our own body and personality, but in everything that lives. This is an insight that, though not correlating with common sense reality for most people, I suspect may one day be proven by geometric principles and mathematical equations and be incorporated uniformly as such in the normative traditions of bioethics, domestic and international law as well as modern physics.

One related hypothesis is that conscience operating collectively within a society spurs normative evolution towards greater coherence with fundamental patterns of symmetry by enlarging our circle of sympathy. If so, then it is likely that rapid global personal communication (for example through the Internet, mobile phones or outputs from an NES project) will facilitate development of such a collective physics-based global conscience by expanding human sympathy towards protecting those global public and environmental goods now critically under threat.

Placing conscience related to global public and environmental good in the formal governance structures of a global NES project won't be an easy task. The dominant governance ideologies of our world – religious fundamentalism, free market corporate globalisation, democratically lassitudinous consumerist hedonism, as we've shown, have little place for environmental sustainability, or the nanotechnology dedicated to achieving it. Corporate globalisation, for example, doesn't assign a value to its largest capital stocks – natural resources and living systems as well as human social and cultural systems.

One of the great boons of a global NES project may be that it assists the debate about the role of new technology in ecological sustainability to be rapidly shifted from the macroeconomic to the corporate organisational level. This will require appropriate regulatory financial incentives (for instance tax rebates, concessions and subsidies, incentives), as well as the development of processes for generating data relevant to suitably refined outcome measures (such as placing ecological impact alongside productivity). More than this it will require an intellectual preparation of the public to accept nanotechnology not as a threat, but as an agent for public and environmental good worth supporting as a matter of individual and collective conscience.

Many international institutions and organisations claim they are constitutionally committed to responding to the demands of what is often termed a global social conscience. Humanity at both national and international levels is beginning to prioritise its environmental as well as public health problems and channel cutting-edge research towards their solutions. Consider in this context, for example, the United Nations *Millennium Development Goals*, and the various international conventions, agreements and understandings that are creating hard and soft norms about issues such as global poverty, loss of biodiversity and environmental degradation.

One way to encourage humanity's collective conscience to support a global NES project is to promote how that project appears to answer (as it must) our greatest needs as well as those of this planet from scientifically coherent first principles. Our Earth, Biosphere 1, for instance has been scientifically proven to provide the following services for free to human society: 1) capture of solar energy and conversion into food, fuel and other raw materials; 2) decomposition and sequestration of organic wastes; 3) maintenance of a favourable gas balance in the atmosphere; 4) recycling nutrients for plant growth; 5) regulation of fresh water; 6) erosion control; 7) generation and maintenance of agricultural soils; 8) control of pests and pollination of crops; 9) a genetic library; 10) limiting ecosystem destruction and regeneration; 11) control of micro and macro climates; and 12) provision of recreation and cultural amenities.

A successful NES project then must appeal to collective conscience by resonating with major survival concerns such as those about population control and ecosystem integrity, as well as the critical need for cheap, readily accessible sources of energy, water and food. To do this its governance structures may need to manifestly reject the paradigm of unlimited economic growth, as well as accept that large-scale behavioural change is necessary but improbable in current circumstances without nanotechnological assistance.

In governance terms, in other words, for a global NES project to succeed, its refocused acceleration of nanotechnology research and development should be widely endorsed as facilitating global public and environmental goods. The concept of global public goods hitherto has been heavily influenced by economics in which discipline it is rooted in the arcane doctrines of public finance and traditionally related to untrammeled access by nation states to the free bounties of nature. It is doubtful whether traditional anthropocentric-focused economics even has a concept of environmental goods.

A definition of public goods more in accord with a global social contract, however, might posit that they involve knowledge, material, infrastructure, or measures (including international legal and ethical norms) that support or fulfil the basic preconditions for human existence; provide benefits from which no individual should be excluded; span national, cultural and generational boundaries; and involve consumption theoretically not creating rivalry or diminishment by use. Environmental goods as a concept would focus on the previously mentioned basic planetary boundaries needed to sustain ecosystems.

Adoption of this approach for the governance foundations of a global NES project would require that key distinctions be made between nanotechnology's role in providing basic public goods (for instance the supply of housing, or clean water and air as bulk societal commodities) and higher-order public goods, which have a more abstract or ethical dimension (such as transparency, accountability, honesty, equity, cost-effectiveness and environmental sustainability in government programs).

Unless the economists are prepared to compromise somewhat on their ownership of the concept, public goods discourse may continue to break down into a strident assertion of one set of societal values or economic ideology against another, rather than becoming an effective analytical tool for global NES project policy-makers.

Thus, the oxygen in our air, the ozone layer, the role of rainforests and oceans in storing CO_2 and the process of photosynthesis as a direct (food and biomass) as well as indirect (coal, oil and natural gas) source of energy are examples of public and environmental goods that standard economic theory appears to have assumed will always be provided abundantly and free by nature to human society.

A global NES project may support global public and environmental goods poorly (if at all) acknowledged by standard economic doctrine. These include democratic input to national and international communication and education systems, technology comparative cost-effectiveness regulatory systems, financial compensation for informants bringing to justice corruption in government and the private sector, reduction of the global arms trade, human trafficking, poverty, environmental degradation, peace, respect for human rights and an equitable international trade regime. Global utilisation of nanotechnology also offers the potential for decentralisation of energy supply, health, education, food and water production, clothing and housing systems.

These may be viewed as novel, emerging conscience-driven constructions of public policy that sustain democratic mechanisms for collective policy development as themselves an important global public good.

Admittedly there is a potential cultural bias and privileging of policy objectives in any nomination of the global public goods or critical public health and environmental problems that a global NES project should address. Each such nomination in effect involves setting that concept in a distinct category of intrinsic worth to society – a judgment with utilitarian and normative aspects. Hence, the important debate about how a global NES project addressing global public goods can also involve discussion of how humanity should optimally order its collective priorities. Identification of certain desired outcomes that a global NES project should address as global public goods is, in effect, also an expression that those outcomes have ethical approbation warranting funding prioritisation in public policy.

Public and environmental good advocacy on behalf of a global NES project undoubtedly will face a range of criticisms from both governments and the corporate sector. The growth of transborder capital investments and offshore tax havens has meant that state sovereignty is no longer solely about defending a physical space, but frequently more about a clique of corporate directors influencing the regulatory apparatus of nation states, through lobbying, the media, political donations and corruption.

As mentioned in the preceding chapter, a global NES project undoubtedly will confront obstacles created by a blurring of respective roles, responsibilities and jurisdictions in the public and private governance spheres (a phenomenon sometimes termed the 'Market-State'). Under the influence of lobbyists promoting monetarist economic ideology, governments, for instance, may become disinterested or powerless to promote a global NES project however scientifically justified it is in relation to public education, public health and environmental protection. The policy leaders of such a 'residual' or 'vestigial' corporate-state are likely to have given up fighting global mass media, as well as well-financed corporate lobbying backed by trade and investment agreements.

To circumvent such problems a global NES project will need to effectively mesh with established international civil society means of supporting individual politicians or governments or private charitable or philanthropic enterprises (such as the Gates, Clinton and Soros Foundations) and whose policies assist global public and environmental goods. Further, nanocomputing as either a primary or secondary focus of a global NES project may allow all citizens to vote on domestic legislation and resolutions of the United Nations.

There may also be a need to link a macroscience NES project with core international law public and environmental goods norms such as the concept of the 'common heritage of humanity' first enunciated in the United Nations Outer Space treaties, *Moon Treaty*, *Law of the Sea Treaty*, and the UNESCO World Cultural and Natural Heritage conventions. The 'common heritage' concept allows for public international law to protect certain public and environmental goods from sovereign exploitation and military use. Whether the same concept also protects such global public goods from corporate misuse is a complex and unresolved issue for positivist legal systems (though not, it is argued, for science-based natural law).

The next section explores how the governance foundations of a global NES project may support the foundational virtues of a global social contract.

4.3 NES FACILITATING VIRTUES IN A GLOBAL SOCIAL CONTRACT

Towards the end of my medical school studies I travelled to Thailand and by a series of coincidences ended up staying at a Buddhist forest monastery in the north of that country on the Mekong River. One of my most memorable experiences was meeting the Ven. Ajahn Thate at Wat Hin Maak Peng. The Ajahn was very old at that time but allowed me to stay in a small wooden hut, study in the temple library and participate in the monks' life including going out begging in the morning for food from the villagers.

I learnt that, historically, the ideal model of Thai society involved young people adopting the life of a monk for a few years to learn inner discipline. This strengthened their bond to those dedicated to Buddhist ideals when they later joined the workforce and started families. The Buddhist monks were seen by the public as embodying the principle that the purpose of human life is to achieve sufficient detachment from the sorrows of this world to alter their consciousness so as to experience the unity of all life, not just in deep meditation but whilst they are awake. Respect for that goal firmed up the social contract as a bond that was fundamentally about developing the highest virtue.

As well as promoting the primary social (but uniquely non-human-centred) virtue of environmental sustainability a macroscience NES project, to have stable global governance foundations, will have to accord with other social virtues that stabilise a global social contract. Many of these have already been discussed. Another illustrative example to emerge with great significance in recent years is inter-generational equity. Like environmental sustainability this social virtue also demands a considerable leap of sympathy away from immediate human concerns.

We have previously explored how such virtues arise and are maintained in societies, just as in individual human beings, through the consistent application of universally applicable principles in the face of obstacles.

A variety of important principles require consistent implementation to develop inter-generational equity as a foundational social virtue of the global social contract endorsing a macroscience NES project. The first holds that the present generation should conserve sufficient natural and cultural resources so that future generations retain an adequate range of options. The second is that the present generation should preserve the quality of the Earth so that it is passed on to future generations in a similar or improved condition. The third posits that the present generation should maintain for future ones a reasonable and equitable right of access to natural and cultural resources.

Justice and equity are other foundational social virtues that must be reinforced by a global NES project social contract. These are traditional social virtues focused on present human interests. Yet there now is a strong emerging environmental component of justice and equity considerations. Policy-makers (including the directors of supranational corporations) support this virtue by increasingly implementing in the face of obstacles the principle that their decisions should acknowledge that the human economy is a dependent subset of the biosphere.

Nanotechnology can be promoted to the public as having the potential to transform corporate globalisation towards the virtues of justice and equity in a global social contract – by making it less dependent on materials-intensive industrial production organised large-scale at particular geographic locations. Domestic nanofactories, however, are a long way from feasible worldwide roll-out. In the meantime it will be important to create governance processes for a global NES project that encourage the corporations involved to discover and disclose their institutional values (including those of their primary stakeholders) and the extent to which they mesh with the social and environmental issues that the rest of humanity has required in its global social contract that such organisations should address.

Legal structures derived from such a reconceptualised understanding of corporations within a global social contract will encourage corporate profits to depend on the needs of citizens wishing to flourish in a sustainable biological future. Their institutional success may be defined for instance in 'sustainability reports' to shareholders in terms not just of traditional economic, but also of community and ecological goals.

It is worth recalling here that EF Schumacher's influential recipe for encouraging large global corporations to locally enhance human freedom, creativity, dignity and environmental sustainability involved organisational structures maximising the opportunities for all workers to develop and promote creative ideas that enhanced profitability whilst doing good works for others and the environment.

If the foundational social virtues of a world social contract manifestly support a global NES project, individuals, households and communities would become more motivated to make corresponding behavioural changes. They might use nanotechnology, for example, to generate the products they need at the place of their use from a much simpler range of universally accessible raw materials, with low biosphere costs, recyclable outputs and benign effluents. Ecosystem health likewise may be encouraged to become as big a research specialty as international medicine or surgery. With a global social contract supporting a macroscience NES project the distinction between developing and developed nations gradually may evaporate as households, communities and nations have their own nanotechnology-based sources of food, energy and water.

Further, if a global social contract promotes a nanotechnology-based economy dedicated to environmental sustainability, cheap labour and growing of basic crops mightn't offer significant global market advantage. Nanotechnology could be valued as assisting poverty to be overcome not only because of the suffering and loss of human dignity it entails, but also because it is a prime cause of environmental degradation.

Perceiving from eternity the issue of establishing appropriate foundational governance structures of a macroscience NES project requires coherence between the underlying global social contract and the principles of science-based natural law. One outcome may be that supranational corporations are explicitly drawn within that social contract as entities presumptively striving to achieve virtue for example by developing nanotechnology (if that is where they decide to invest) for human and environmental good in appropriate proportion with shareholder profits. Community involvement in the corporations that shape global nanotechnology governance likewise becomes an important normative extension of commitment to core foundational social contract virtues such as environmental sustainability.

In the passing era of corporate globalisation the specific geographic, political and legal locality in which people physically lived became less significant to their employment, income, well-being and culture, as well as their political allegiances and the virtues they hoped to develop in life. Raw materials were grown or mined, then shipped to distant locations where cheap labour allowed low-cost manufacture of food, clothing and household products to be shipped back hundreds, even thousands of miles to satisfy non-essential needs promoted by corporate advertising. It was a process unlikely to be coherent with rational models of a global social contract. As citizens embrace nanotechnology-based environmental sustainability as a chief virtue in their social contract 'Think Globally, Act Locally' will become a valued lifestyle available equitably to the world's population including its corporate members.

Social contract-related criticism that establishment of a global NES project may face include: 'isn't it unjust or inequitable to focus on nanotechnology-based environmental sustainability when the world has equally pressing and apparently distinct problems such as pandemics, international terrorism, nuclear weapons proliferation, overpopulation, pollution and man-made climate change? Aren't you simply playing God in seeking to manipulate nature this way? All you're doing is looking for a technological fix?'

On the science-based natural law approach advocated here, however, whenever a technological fix has become possible then the relevant problem appears to have fulfilled its duty. Just think how much ingenuity and human determination to resolve complex problems we take for granted when we switch on an electric light, or turn a tap. What aspect of our political or religious systems works with as much consistency and predictability as such technology? Is this a gap in efficiency we simply have to accept? Technology can be regarded as a hardening in time and space of the social virtues that allowed it to develop. In this sense it becomes a record of a society's virtues long after that society has perished.

A global NES project thus will have a long-term symbolic importance for our generation. It will demonstrate to future generations exactly what foundational virtues we were committed to as a species. As such this project should explicitly involve adaptive governance processes that seek to maintain virtues such as resilience, the capacity to respond efficiently to changing scientific evidence of public health and environmental need, as well as to foster involvement with civil society networks and industry.

The underlying global social contract must be manifestly clear that such processes do not involve privileging any one comprehensive moral doctrine or approach to institutional success. They must demonstrate that the global NES project is not seeking to impose a single version of a solution, but is encouraging a multiplicity of approaches through a wide variety of governance arrangements capable of supporting universal values, without unduly antagonising particular cultures or established corporate, political and religious interest groups.

Nanotechnology, the citizens of the world will hopefully appreciate from the activities of such a global NES project, is capable of assisting our political and cultural allegiances to embrace ecosystems within our expanded circle of ethical concern. This will link them together in the mutual exercise of their free will under a rule of law in which the biosphere is accorded not merely instrumental, but intrinsic value under a global social contract.

4.4 SUMMARISING NES PROJECT GOVERNANCE CRITERIA

As a lawyer and medical student I used to study and teach yoga. I wasn't much good at the latter, but did publish a paper in *Medical Hypotheses* trying to explain the strange yogic practice of early-morning auto-urine drinking as physiologically based on deconjugation of urinary-excreted melatonin to its active form in gastric acid.

One of the most peculiar things I came across while reading yogic literature was the number of esteemed yogis who claimed that their meditative practices had allowed them to levitate, roll space into a tube to see at vast distances and so forth. Similar claims exist amongst the traditions for instance of Christian saints, Muslim sufis and aboriginal shamans.

One could of course dismiss such statements as unashamed fictions produced by idolising followers to perpetuate religious authority and power, or the misapprehensions of gullible, naïve or over-excited minds. But why not assume that such mystic activities are genuine anomalies in the natural order of things, equivalent in other words to the constancy in the speed of light regardless of its source?

Such mystic anomalies seemed to be acquired through prolonged meditation and fixed concentration of mind. Indeed, texts such as the Yoga Sutras of Patanjali set out a program of developing virtue as a precondition to perfecting fixed voluntary concentration of the mind, in which these mystic powers are claimed to routinely appear as unwanted side effects. What happens to the pre-existing (non-experienced-based) three dimensions of space and one of time in a mind that is able to develop fixed concentration? Isn't this the type of question that brings natural law into coherence with modern physics?

The point being made is that governance systems like that of a global NES project focused on science-based natural law, like science, are unlikely to advance unless they manifest a willingness to explore anomalies that don't appear to correlate with common sense, but do appear to involve humanity's 'higher' consciousness.

Many years ago, the economist Galbraith in *The Affluent Society* warned of two great dangers for a society characterised by affluence that are equally applicable to one dominated ideologically and practically by nano-technology. The first is that such a technologically provided-for community will settle into a comfortable disregard for those it excludes and the second is that it will develop a political doctrine to justify that neglect. Designing appropriate governance structures for a global NES project can be viewed as an antidote to these twin social poisons.

Let's summarise then some of the arguments made to this point about the reasons for, and foundational governance structure of, a global NES project. Consideration of how ethical and legal norms are created and supported suggests such a project must have a funded segment dedicated to promoting informed global public debate on its ethical, legal and social implications. That funding will draw to it a range of ethical and legal experts who (as they did with the ethical, law and social implications (ELSI) component of the Human Genome Project) will provide an academic flak jacket for instance protecting the project and its supporters from irrational mass media attacks.

The stakes are even higher for a global NES project than they were for the Human Genome Project. Humanity is facing unprecedented challenges to its survival that a global NES project must promptly and directly address. A significant majority of the world's most eminent scientists, for instance, after investigating the issue have reached consensus that man-made climate change must be tackled immediately, that deforestation, population growth, extinction of species, pollution and destruction of ecosystems must be curbed.

The case has been advanced in this chapter in particular that in order to rise to such a challenge a global NES project must have governance arrangements that are underpinned by ethical and legal norms coherent with foundational social virtues such as justice, equity and respect for environmental sustainability. Further, the point has been made that although it seems to be common sense that governments of nation states create such legal norms, in actual fact they arise as attempted representations of symmetry and harmony manifesting as physical laws. For a global NES project to succeed, its governance framework must be established under a social contract that ensures policy-makers are unlikely to distrust the science involved, that it is not lobbied against by corporate representatives striving to maintain profits and that the idea is efficiently embraced worldwide by human communities, households and individuals.

It follows that one of the chief governance criteria of a global NES project must be that the aspect of nanotechnology it centrally focuses upon remains environmentally friendly throughout its life cycle. Nanotechnological solutions that poison the environment or degrade our society will be of little relevant benefit, no matter how brilliantly conceived, or how great their superficial attraction to those in charge of mass production and sale of consumer goods.

A global NES project also should in a powerful symbolic way facilitate our capacity to conceive long-term solutions to the great public health and environmental challenges of our times. As the economist and Quaker Kenneth Boulding once stated, the welfare of the individual depends on the extent to which he or she can identify himself with others, and the most satisfactory such identification is with a community flourishing not only in a particular geographical space or spaces, but also into the future.

Likewise, according to science-based natural law as manifested in a global social contract, one of the criteria for selecting a global NES project should require the selected nanotechnology innovation to provide ordinary people everywhere with the means of doing profitable and intrinsically significant work. Its governance framework could incorporate an ethical requirement that the cost of producing the selected nanotechnology products conforms equitably against the level of incomes in the society where they are to be used. To the extent that large-scale factories for making nanotechnology initially exist (they are likely to be replaced by nanofactories), such a requirement could specify that the upper limit of capital investment in that workplace implementing the products of a global NES project be linked proportionally to the annual incomes of its employees.

In terms of symbolic importance, a major task of policy-makers in preparing the governance framework of a macroscience NES project will not be to signal a higher purpose that all future generations must embrace. Rather, its role should more be to highlight how so many of our preoccupations are most dangerous in terms of the survival of our society and that of our ecosystem (particularly the making of money through the production and purchase of private consumer goods) are compelled by traditions and myths that may be inaccurate, out of date and dysfunctional. Humans considering embracing a global NES project should readily be able to see evidence that it aims to make them free to survey other, more aspirational, long-term or 'deep future' options as regards the meaning of their existence.

The governance arrangements of a global NES project, as part of the resource allocation issue involved in prioritising public funds for it, cannot escape addressing whether the core of economic stabilisation for most states should continue as an alternative to be a large military expenditure sustained by taxation of citizens. To garner widespread public support such a project should also manifestly work to governance criteria designed to ensure our most brilliant minds and the research apparatus supporting them are not thereby directed to enhancing consumer goods, corporate profits or military technologies. A macroscience NES project should obviously be predicated on leading future generations to escape poverty and physical insecurity.

The governance arrangements forming the conceptual bedrock of a global NES project should allow the public readily to mesh nanotechnology into what Bertrand Russell termed humanity's 'profound instinctive union with the stream of life'. Examples of such global NES governance principles could require domestic governments and regulators assessing the resultant nantechnology products to: 1) implement a precautionary approach to related public health and environmental risks; 2) ensure uniformity of mandatory nano-specific safety regulations protecting the health and safety of the public and workers; 3) facilitate virtues such as transparency, public participation, justice, equity and environmental sustainability; 4) facilitate manufacturer requirements for clear labeling and liability in accordance with the rule of law; and 5) ensure NES project governance arrangements conform to key international statements of bioethics and human rights such as the already mentioned UNESCO UDBHR and the United Nations ICESCR.

Democratisation of a global NES project could be enhanced by governance measures that, for example, allow all adult, competent citizens of every nation state, say every 50 years, to refresh by a free and fair vote their support for the basic provisions of not just their constitution (including the human and environmental rights protections therein) but also the basic structure of a global social contract emphasising the United Nations and its political and humanitarian organs but also the non-governmental organisations of international civil society.

Likewise worthy of consideration are governance mechanisms that permit every piece of domestic legislation, or United Nations General Assembly or Security Council Resolution to be subject to online Internet voting by all registered citizens. Each of these options will be enhanced by global nanotechnology-based communications systems, but require a stronger commitment to the basic science-based natural law norm requiring democratic involvement as a precondition for legal legitimacy.

Whether or not a global NES project ends up having its core governance arrangements specified in a United Nations convention (a multilateral treaty binding on nation states under international law) its core vision must be embraced enthusiastically in the conscience of the peoples of the world. Its governance principles must be seen explicitly (that is in some formally published document) to support primary social virtues such as justice, equity and environmental sustainability.

In the following chapters we examine against the criteria explicated here some of the main candidate areas for a global NES project.

5. Nanotechnology for sustainable food, water and housing

> The social structure of agriculture, which has been produced by –
> and is generally held to obtain its justification from –
> large scale mechanization and heavy chemicalisation,
> makes it impossible to keep man in real touch with living nature;
> in fact, it supports all the most dangerous modern tendencies
> of violence, alienation, and environmental destruction.
> Health, beauty and permanence are hardly even respectable subjects for discussion.
> – EF Schumacher, *Small is Beautiful*

5.1 NES PROJECT ON GLOBAL FOOD SUPPLY?

When our family go shopping for food, we prefer to walk around farmers' markets than down the aisles of large-scale commercial shopping centres. Farmers' markets are like community fetes. There are a lot of associated communal activities going on – raffles for charities, games for children, people busking or selling art. You get to talk to the person who grew the food being sold about their philosophy and values. The farmer who grows our beef, for example, herds his rare Belted Galloway cattle without dogs by calling out for them. The vegetable growers tell of the tribulations they have to go through to be certified organic. You walk around in the open air surrounded by the smells of real farm animals, with freshly picked produce still flecked with dirt. This is a clear distinction from the sanitised space prioritising advertising, air conditioning and muzak that characterises the large commercial supermarket and the values of its owners. What you lose in rock bottom price, consumer choice, artificial flavourings and preservatives at a farmers' market, you make up for in relationships with people and the natural world.

Given the level of scientific and industrial interest, there is no doubt that food and agriculture could be a strong candidate for a global NES project. Some of the supranational food companies with major investments in nanotechnology research include Altria (Kraft Foods), Cadbury Schweppes, Campbell Soup, HJ Heinz, Nestle, Pepsi and Sara Lee. They are working on developing nano-enhanced foods that can change their colour, flavour or nutritional properties according to a person's dietary needs, allergies or taste preferences. Large agrochemical food companies such as DuPont Food Industry Solutions, Goodman Fielder and Group Danone also have significant nanotechnology research components looking, for example, at nanosilver in packaging to reduce bacterial contamination and prolong shelf life, as well as incorporating otherwise unpalatable additional nutrition into popular food items.

Silicate nanoparticles (for instance, in plastic packaging or wax-like coatings) have been developed to slow down oxygen, moisture and bacteria penetration and extend food shelf life, while nanocoating sprays are being used to preserve fruits and vegetables. Nanocomposites can provide pathogen, gas and oxygen barriers facilitating long-term food storage without refrigeration and reducing packaging waste through use of monolayer films. Storage bins are being lined with antibacterial and antiviral nanosilver or to create packages that will change colour if their food contents go off.

Nanotechnology may also provide food deterioration sensors (such as light activated O_2 sensing ink) to control the role of ripening of fruits, monitoring of food temperature, moisture and size. Nanotechnology food wraps are being produced to detect contamination and biodegrade when discarded, or utilise barrier properties from tortuous paths in the coating. Nanotechnology-based radio frequency ID tags can be placed inside food or food packaging to not only facilitate scanning at the counter, but potentially check up on what happens to the food after it leaves the store.

Nanoparticles and nano-emulsions likewise are being utilised to enhance food desirability, for example, by ensuring easy pouring, improved flavour and texture, or by delivering a broader range of nutrients more efficiently and with better traceability. Nanotechnology coverings are being designed to enhance artificial flavourings and colourings, reduce fat, carbohydrate, or calories from foods, as well as improve their protein, fibre or vitamin content and smell. Drink bottles are made with nanoparticles embedded in the lining to stop carbon dioxide leaking out of the bottle.

There is little doubt, however, that nanotechnology developments in food can also more centrally address issues of justice and equity under a global social contract. As such, food security is a plausible topic for a global NES project according to the criteria previously mentioned. Nanotechnology, for example, may assist in enhancing the nutritional value of many subsistence food crops and products. One way it can do this is by increasing the amount and bioavailability of otherwise unpalatable or undigestable nutrients in culturally standard dietary components or food aid. Methods include nanoencapsulation to enhance nutrient absorption by the body.

Nanoparticles, similarly, can be designed to absorb the vitamins in produce such as orange juice, where the vitamin C levels deteriorate quickly after the fruit has been juiced, and release them upon consumption.

Nanotechnology, relevantly to a potential global NES project, also is being developed to enhance food security by assisting disease detection and treatment in livestock and plants. Metal and metal oxide nanomaterials such as Ag, Fe, ZnO and magnetite (Fe_3O_4) nanoparticles are being utilised to detect heavy metal ions and volatile organic compounds in water and soils.

A global NES food project might assist to develop more potent fertilisers (for example developing methods for enhancing nitrogen fixation and dissolving more readily in water for ease of application) or to facilitate their release at the right time for crops. Such a project could create specifically targeted pesticides, for instance in response to specific detected threats from drought, flood pests or pollutants. Nanosensors embedded in fields and farm animals likewise might be developed to detect disease before it infects the whole crop or herd, as well as to allow machinery to deliver nutrients, water or fertilisers for maximal effective impact. There are undoubted public good and public health benefits from such a global NES food project that would satisfy the previously discussed foundational virtues and principles of a global social contract.

A global NES food security project would support developmental and environmental sustainability by facilitating the equitable long-term availability, access and utilisation of food for the entire populace. Additional benefits of such a global NES food project might include a reduction in food prices for local consumers, improved nutritional health associated with greater variation in diet and a dispersal of risks associated with nationally centralised food production in the face of natural disasters such as drought or flooding. A global NES food security project, by encouraging developing nations to focus away from export sales of cash crops, could begin to redress malnutrition and starvation, particularly amongst the urban slum-dwelling poor, refugees and small-scale farmers who cannot compete with the low prices, mechanisation and distribution networks of international agribusiness.

A global NES food security project thus would address significant problems related to global public health and environmental sustainability that are already a major focus for international institutions. It is estimated that as human population exceeds 9 billion there will be a doubling in our demand upon food production worldwide by 2050.

Malthus argued (in his essay 'On the Principle of Population' in 1798) that population, when unchecked, increases in a geometrical ratio; but that subsistence grows only in an arithmetical ratio. This conclusion was based on two postulates that he considered natural laws. The first was that food grown in fields by then known processes would always be necessary to the existence of man and the second was that the relationship between sexual passion and birth of children would inevitably be constant. Malthus's essay was partly a polemic against the social anarchism promoted by visionaries such as William Godwin (Mary Shelley's father) who rejected any social system dependent on government and promoted a simplified and decentralised society with a dwindling minimum of authority, based on a voluntary sharing of material goods.

Yet, Malthus failed to foresee how his formulations of these natural laws would be undercut by technological developments such as the Haber-Bosch process (allowing the fixation of nitrogen in ammonia to make the fertilisers that support much of the world's agriculture), the oral female contraceptive pill, the inverse relationship between education and population growth and modern methods of artificial reproduction. Similarly, if for example a global NES project facilitates widespread use of nanofactories that make food from basic ingredients such as sunlight, water, carbon dioxide and common minerals then global food production (contrary to Malthus's formulation of natural laws) could increase in a geometrical ratio. Second, nanobiotechnology may reduce the human necessity for such large amounts of food or energy and water intensive agriculture to accord adequate nutrition for each person – either through its impact on the required agricultural processes or our metabolic requirements.

Nonetheless, citizens across the Earth thinking about whether to support a global NES project focused on food and agriculture might see nanotechnology as an expensive threat to their safety, livelihoods and quality of life. When farmers simply have no land, no security, water, seeds or fertiliser to farm it, then resource priorities seem not to suggest nanotechnology as the solution. Far more immediate answers may appear to reside, for example, in the ethical and human rights-based governance efforts that support policies such as encouraging people to buy 'Fair Trade' labelled products, grow their own vegetables in suburban plots, and support local food producers as well as the right of local communities to control their local food trade while avoiding the practical necessity to eat highly processed, packaged, artificially preserved and sweetened foods.

A global NES food project, to garner widespread public support, would thus have to explicitly link with governance efforts promoting organic agriculture including ecologically sustainable practices that involve non-synthetic nutrient cycling, exclude synthetic pesticides, regenerate oil quality and do not involve genetically modified organisms.

Over 100 nations currently have some organic agriculture capacity (chiefly serving urban dwellers) with more than 70 million acres under organic management globally. Supporting international organisations include the International Federation of Organic Agriculture Movements (IFOAM). Without careful management a global NES food project is likely to be viewed as a threat by this sector. IFOAM has produced recommendations tending in this direction. These state in part:

> Governments should establish comprehensive and precautionary legislation to manage the risks associated with nanotechnology and should safeguard the right of producers to refuse to use it and the right of consumers to say no to nanofoods and nanomaterials in other products. Nanomaterials must be considered as new substances even if the properties of their larger scale counterparts are well known. In the absence of a moratorium, all food, agricultural and other products which include manufactured nanomaterials must be clearly labelled to allow consumers to make an informed choice. Those who seek to commercialise manufactured nanomaterials should be required to demonstrate the safety of the new technology before the technology is released. If health and safety risks become evident, the owners of the technology should be strictly liable for the damages caused, including but not limited to the losses related to the contamination of organic food and fibre.

Let's look in more detail at some of the safety issues a global NES food project is likely to face. There is no doubt that ensuring the safety of food containing nanoparticles will be a major public concern about whether to support such a project. Consumer resistance and environmental protest caused major problems for the genetically modified food industry (although the power of these groups to influence governance was enhanced by business conflicts about loss of premium markets, high costs of segregation and handling).

An NES food project's governance initiatives here could, for example, include funding to develop a core of research and adequately trained academics in multiple disciplines able to effectively respond to such concerns in the media. It could also involve a centralised source of information readily accessible to the public on nanotechnologies used in the food sector, as well as lists or formal monitoring of nanotechnologies being researched worldwide by companies working in the food sector.

Such a global governance initiative additionally might encourage debate on issues such as whether food labeling should be based on nanomaterials present in products, and not on the process by which they were manufactured or farmed. It might consider whether such labeling should include only information a citizen is presumed to need to use the product safely, rather than on a 'right to know' basis as was the case in Europe, for example, when the precautionary principle was applied to the use of genetically modified organisms in food.

A related requirement in the governance arrangements of such a global NES food project might be that manufacturers of agricultural produce containing (or packaged in) nanotechnology prepare a public dossier of safety tests to assist pre-market regulatory safety review and approval as well as challenge by civil society organisations. The governance components of such a project might also examine whether for regulatory purposes any definition of 'nanomaterials in food' should exclude those created from natural food substances or focus exclusively on food and agriculture nanomaterials deliberately engineered to take advantage of their nanoscale properties.

The United Nations through one of its specialised agencies, in collaboration with relevant stakeholders, could assist a food and agriculture global NES project by supporting the development of voluntary codes of conduct and uniform domestic legislation designed to protect public health and environment through appropriate research and monitoring of the life cycle of nanoparticles in food. Such a process might be an extension of the United Nations *Globally Harmonized System of Classification and Labelling of Chemicals* (GHS). A case-by-case approach would allow the responsible development of low-risk nanotech food products where safety data are available and a selective moratorium on products where it is not.

Standards could be developed as part of this global NES food project for detecting and measuring the type and level of nanomaterials in produce. The WTO could be involved on the basis that ensuring this research results in practical tests used by enforcement agents will be an important step in securing the safety and security of food imports. The same mechanism might produce and maintain an accessible list of publicly available food and food packaging products containing nanomaterials, as well as a mandatory confidential database of nanomaterials being researched by the global food sector.

Such a global NES food project would have to deal with issues such as public demand for regulatory standards preventing companies 'knowingly' placing unsafe food on the market. This standard can result in companies seeking to retard the publication of data that compromise their products' safety. Another safety issue to be addressed is that increasing amounts of food globally are likely to contain nanoparticles unintentionally as a result of wear of food processing machines and environmental pollution. The project additionally would need to review whether new nanoscale materials should be viewed as 'novel ingredients', while familiar materials that had been engineered to the nanoscale are covered under existing regulations.

Yet most importantly a macroscience NES food project, to achieve the necessary widespread public and policy acceptance by satisfying the governance criteria outlined in the earlier chapters, must manifestly accord with the global social contract and science-based natural law. This means that social virtues such as justice, equity and environmental sustainability should clearly appear to the global public as central to the principles underpinning its operations.

Practical examples of this approach would include explicit governance mechanisms aiming to reduce production and consumption of environmentally damaging animal source products. Other policies likewise might support rural development and food production techniques that promote ecological sustainability and encourage food system diversification including investment in urban agriculture.

A global NES food project's governance processes similarly could promote nanotechnology's role in preventing urban healthy food 'deserts' and 'nature deprivation syndrome' in children, halting collapse of fisheries from pollution and over-catching and loss of prime agricultural land to threats such as urban sprawl or coal-seam gas fracturing.

Such processes in similar manner could restrain excessive corporate ownership of nanotechnologically enhanced food through patented seeds and genetically modified crops as well as food transport and sale monopolies. Presently, over half the world's population undertakes agricultural production and contests with a handful of global food corporations whether the concept of sustainability in this sector requires small-scale methods, local systems, organic farming methods and preservation of biodiversity and cultural traditions, or high yield, pest-resistant and stress-tolerant genetically modified and patented crops.

For instance, as the climate change 'tipping point' is approached, with global temperatures and unpredictable severe weather events increasing (particularly as a result of increased water vapour in the atmosphere), a small number of supranational plant breeding companies may trawl through gene banks searching for heat-tolerant plant genes to be genetically engineered (potentially using nanotechnology) and patented for mass production in the major growing zones. The crops could have terminator (sterile seed) genes designed to enhance profits (by requiring more purchases) but marketed as a mechanism to protect against contamination. Although large-scale corporate investment will be required for the success of an NES food project, without careful prior attention to governance arrangements it could result in global food security depending on the unpredictable goodwill of a small number of supranational agribusinesses.

Similarly, the success of many touted nanotechnology breakthroughs may assist multinational food globalisation companies to resist community initiatives such as the slow and organic food movements. A corporate-dominated global NES food project, though it may help to grow corporate shareholder profits and director incomes from consumer sales, because of the orientation of those companies may not increase the accessibility of global populations to nutritionally valuable food. For instance, the use by corporate food multinationals of nanotechnology in food packaging to prevent spoilage may only exacerbate the social and environmental problems associated with transporting and marketing food over the long times and distances implicit in the corporate globalisation of food model.

The global public similarly are entitled to be concerned that a pre-eminent dedication of nanotechnology through such a project to the task of remediating polluted agricultural soils and waters will merely leave untouched for longer the causes of such environmental problems.

Just as increasing numbers of informed and concerned citizens globally are beginning to favour locally produced organic food, nanotechnology might well entrench reliance on chemical, and energy and technology intensive, corporate owned agricultural businesses, promoting the transport of food over even greater distances. A global NES food project thus might unwittingly continue a pattern of large-scale, genetically uniform crops, itself creating a threat to food security via a reduced biodiversity and compromised resistance to pests.

To take another example, synthetic biology could use nanotechnology to improve plant production by improving photosynthetic pathways. This could alleviate one of the major problems with accelerated climate change – famine. But even if these advances progress, do we have the domestic and international regulatory systems in place (through for example WHO, Oxfam, the United Nations) that are likely to transfer this nano-enhanced food to where it is needed most? What will happen to crops of the developing world such as vanilla, rubber, cocoa, coffee, cotton, maize, when nanotechnology allows replacement foods to be coaxed from cell cultures in large laboratories?

Further, even if all the promise of nanotechnology in food and agriculture is realised, it is likely to result in the foreseeable future in only a qualitative improvement in crop yields and nutritional value. These will be valuable (particularly to people living in poverty), but may be unduly expensive and involve significant public health and environmental concerns (particularly in the use, for example, of ecosystem-damaging nanosilver in food packaging).

In one best-case scenario, a global NES food project could facilitate a population shift from urban manufacturing to rural agriculture and community-oriented crafts and guilds in the countryside. In a world committed to the primary social virtue of NES, locality and interaction with nature would become more important to our business models and lifestyles, even as global communication makes it less critically important to our communication and our political allegiances. Yet to achieve such broad-ranging benefits a global NES project would have to not only resolve the issues of corporate involvement in governance, but also tackle the problem of global water supply. We turn to the latter issue in the next section.

5.2   NES PROJECT ON GLOBAL WATER SUPPLY?

We live in Canberra, Australia, being extremely fortunate to experience droughts as only involving water restrictions that limit irrigation of gardens, length of showers or cleaning cars. At such times, people here don't yet die of thirst or from infections through being forced to consume contaminated water acquired in situations of great stress. Our drinking water is normally very clean – running from creeks in a pristine catchment area (the Brindabella Mountains) into dams of increasing size. One of the great pleasures of bushwalking in those mountains is to be able to drink from streams uncontaminated by fertiliser, pollution or waste.

Recently, however, the town of Goulburn just 50 miles to the north of Canberra almost ran out of drinking water during a prolonged drought, something that had never happened before. In a nearby area of Australia scarcity of water is creating profitable opportunities for supranational companies to buy and sell water rights along a large river system known as the Murray–Darling Basin. Large Australian coastal cities such as Sydney are now installing old photosynthesis-fuelled desalination plants and many outback regions rely on tapping a huge aquifer the level of which is dropping in many areas directly as a result of tapping by corporations who wish to sell its water in plastic bottles (I recently successfully ran a campaign to eliminate bottled water from my college at the university owing to dangerous chemical leaching from the low-grade plastics involved). This aquifer also is being subjected to contamination through chemicals released by explosive coal-seam 'fracking' organised by corporations determined to profit from natural gas.

Seventy per cent of the world's fresh water goes to agriculture and the common sources are drying up. Aquifers are being depleted or degraded with rising salt and falling groundwater levels. Glaciers are shrinking and river flows diminishing. Desalination plants consume vast amounts of old photosynthesis fuel. Other problems include pollution of groundwater owing to inadequate sanitation systems, algal blooms fertilised by the phosphorus and nitrogen contained in human and animal wastes, detergents, fertilisers, pesticides, heavy metals, as well as salinity caused by inefficient agricultural irrigation.

Yet such problems with water supply (important as they are) are insignificant alongside those of the 2 billion people in the world today (mostly in developing nations) who either lack access to adequate clean drinking water, or go hungry through inability to irrigate their crops. Globally, millions of people are forced by necessity to drink and cook with polluted water, or travel significant distance through many hazards to acquire it.

The United Nations *Millennium Development Goals* require that the number of citizens without sustainable access to safe drinking water and basic sanitation be halved within 15 years. Water is already the world's third largest industry, behind only oil and electricity. The scarcity of potable water will increase with global climate change, but the infrastructure for providing equitable access to it is outdated in most countries. As a scarce but precious commodity fresh water is likely to be the subject of significant trade disputes.

What case can be made then for a global NES project focused on improving equitable access to water? Nanotechnology does appear capable of resolving some major problems associated with global water supply. One illustrative example involves the removal from groundwater of arsenic accumulated above WHO safety levels as a result of natural leaching from bedrock or from industrial effluent. At Rice University in the USA 12 nm magnetic iron oxide nanoparticles have been developed that have been proven to bind over 99 per cent of the arsenic in the water. By applying a magnetic field, the nanoparticles and their attached arsenic can be separated and simply removed from the water.

Another example involves use of nanotechnology to degrade trichlorethelyne (TCE), a common feedstock chemical, dry cleaning chemical and constituent of solvents, adhesives and paints. TCE is carcinogenic, causing liver and kidney damage and birth defects. It is a common contaminant of groundwater. But iron nanoparticles (with trace amounts of palladium in a slurry) can break down TCE in groundwater. A similar nanotechnology-based mechanism can clear groundwater of chlorinated organic solvents, pesticides, PCBs and some heavy metals.

Similarly, alumina nanofibres can reduce fouling of water by providing an electropositive surface that repels clogging agents. The ancient Romans used to use the non-engineered nanoscale components of clays for water purification. Nanoclay/zeolite membranes, likewise, are highly chemical, mechanical and temperature resistant, and purify water by utilising a large surface area and absorption capacity. In another application, nanoscale lanthanum oxycarbonate particles bind to the phosphates that provide nutrition for algal blooms thereby potentially reducing their potentially deleterious impacts on fresh water creeks and rivers as sources of potable water.

An emulsion of nanospherezymes can reduce fouling on filter surfaces. Furenomes in nanofibres can give anti-microbial properties to filters.

Nanoporous zeolites, polymers, attapulgite clays and magnetic nanoparticles can also be used for water purification, treatment and remediation. TiO_2 nanoparticles can assist the catalytic degradation of water pollutants. Nanotechnology also may allow improved water purity sensors. Non-fouling water pipes, in effect pipes that never get wet, can be developed using nanotechnology. Such a superhydrophobic lotus-leaf-like surface is self-cleaning as dirt only loosely adheres to the surface.

Yet, it is not all that clear that the global nanotechnology industry would be willing to invest in a global NES project focused on such issues. Instead (to take a pertinent example) washing machines are advertised across the world as so much more efficient because they contain a system that releases nanosilver particles and ions into the wash. This destroys the biofilm (a coating of little bugs, fungi and bacteria) that builds up inside on the metal. Corporations place nanosilver in washing machines because its results seem likely to be attractive to consumers – making clothes feel fresher and smell sweeter. Yet such widespread use of this nanoparticle will create significant problems for environmental sustainability. Indeed, nanosilver for water purification is a good global NES water security project case study to focus on in more detail, if only because it is the most widely used nanoparticle in consumer products at present. Further, influential regulators such as the US Environmental Protection Agency (EPA) have done little more than require manufacturers who use nanosilver in products to voluntarily produce data. Few companies have taken the time or expense to do this.

Nanosilver in waste water was listed in 2010 by a team of public health experts as one of 15 nascent issues that could deleteriously affect the conservation of biological diversity. While small amounts of silver do exist in streams and oceans, the concentrations are very low. Studies seemed to show that the type of nanosilver particles used in these washing machines is much more toxic to bacteria than normal silver. Nanosilver, like mercury, can accumulate in edible sea creatures like oysters, shrimp, crabs and lobsters. Indeed, once released into the environment nanosilver is accumulated but not degraded by organisms.

Many studies have attempted to compare the toxicity of nanosilver with silver ions. In one experiment where 10–20 nm nanosilver particles in solution were applied to nitrifying bacteria, nanosilver particles less than 5 nm showed a greater inhibition of nitrification than silver ions (Ag^+). The degree of inhibition strongly correlated with measured intracellular reactive oxygen species (ROS) concentration for both nanosilver and Ag^+; though at the same total Ag concentrations Ag^+ generated less ROS than nanosilver and in both cases ROS generation was inhibited by absence of sunlight. Even when the phenomenon of nanoparticle aggregation in biological systems was taken into account, such results supported the finding that nanosilver (particularly at the lower edge of the nano-range) may present a unique and dangerous toxicity even at low concentrations in the environment.

Similarly, it has been found that nanosilver (determined by transmission electron microscopy to be of 20 nm mean particle size) appeared to have a worse impact on gene expression than silver ions (in the form of aqueous $AgNO_3$) when exposed in de-ionised water to *Caenorhabditis elegans* (a soil nematode and the first multicellular organism to have its genome completely sequenced).

Nanosilver in water purification systems also appears to have a major deleterious impact on photosynthesis. The role of Ag^+ in determining the toxicity (in terms of photosynthetic yield) of 5 and 10 μM nanosilver suspensions has been assessed in freshwater algae (*Chlamydomonas reinhardtii*) in the presence of the Ag^+ ligand cysteine. Inhibition of photosynthesis by nanosilver over one hour was similar (60 per cent) at cysteine concentrations between 10 and 100 nM, with a complete abolition of photosynthesis at an equimolar concentration of cysteine.

When capacity to inhibit photosynthesis relative to control was related to the Ag^+ concentration after two hours, nanosilver appeared to be more toxic than the ion source $AgNO_3$. This research supports the view that nanosilver, like silver itself, owes a significant component of both its bactericidal effects and toxicity to the rate of release of free silver ions. The latter finding suggests that nanosilver may provide a uniquely dangerous slow-release mechanism allowing toxic silver ions with high *in vitro* toxicity for aquatic organisms to environmentally persist.

The problem of nanosilver's environmental toxicity is already acute, regardless of the potential influence of a global NES project in this area. The volume of environmentally deposited waste containing residue nanosilver, for example, is increasing proportionally with its utilisation in domestic products and medicinal applications. Nanosilver released in domestic waste water may have a variety of risk-laden fates, including being converted into ionic silver, complexing with other ions, molecules, or molecular groups, agglomerating, or remaining in nanoparticle form. Its biocidal and catalytic effects adversely impact on a wide range of organisms including bacteria, fungi and earthworms in soil, through reaction with other toxic substances, its pollution of groundwater and accumulation along the food chain. Waste water treatment relies on heterotrophic microorganisms for organic and nutrient removal, while autotrophic microorganisms play an important role in nitrification. Nitrifying bacteria in sewerage systems are especially susceptible to inhibition by silver nanoparticles.

A global NES water project that focused on the purifying role of nanosilver, of course, could ensure significant funding for research with long exposure times, covering different estuarine environments as well as investigating the effects of nanosilver in fresh water systems. This could be designed to acquire information about mass loading in the environment for the purposes of nanosilver risk assessments, government reporting requirements and manufacturer product information.

If nanosilver becomes a significant component of a global NES water supply project, then that project will be unlikely to garner widespread scientific or public support. For a start, nanosilver will have a notorious reputation environmentally from being present in a lot of other products that when disposed of will leach that substance into the waterways: food packaging, fridges, vacuum cleaners, air and water filters, medical waste such as wound dressings, paints, surfaces in public places such as toilets, clothes, shoes and socks, toothbrushes, soaps and hair products, deodorants, bedding, baby bottles, toys and wipes. Amongst other deleterious outcomes might be more allergies in children (as they are denied exposure to a range of bacteria) and also the promotion of bacteria that are more resistant to antibiotics. Focusing a global NES water project on interventions such as nanosilver would be unlikely to satisfy social contract governance requirements specifying, for example, concern about future generations and environmental sustainability.

In summary, there can be little doubt that a global NES project focused on improving water quality worldwide (particularly through the already widely marketed anti-microbial nanosilver) could achieve goals coherent with the foundational social virtues of a global social contract. Nonetheless, it would not immediately address issues related to equity of global food and energy supply and would involve significant long-term environmental toxicity risks that undercut those virtues.

5.3 NES PROJECT ON GLOBAL HOUSING AND URBAN DESIGN?

In my twenties I used to travel to India to study yoga and to Thailand to study Buddhism. Such trips required that I pass (as rapidly as possible) through two very overcrowded and poorly designed cities – Calcutta and Bangkok. It was always refreshing to leave the environmental degradation and moral complexities of such megacities for rural areas that, despite their poverty and overcrowding, were often characterised by greater simplicity of living and open-heartedness.

New technology in the form of the handmill, Karl Marx argued in *Capital*, made possible the power of the feudal lord and the steam mill bequeathed to the industrial capitalist. In our own age, the factory, container ship and the cargo plane in large part gave us corporate globalisation with executive salaries out of rational proportion to the trickle down benefits of this process to remedying global poverty. What new governance structures likewise will be introduced to the world when it involves ubiquitous use of nanotechnology in its global housing and urban design?

Imagine Calcutta in the twenty-fifth century CE after the conclusion of one model of a global NES housing and urban design project. There may be cars not powered by petrol. Trees and plants may grow along all the streets, for the aesthetics, not for need as nanotechnology now allows each house to perform many of the environmental tasks that used to be the responsibility of trees. No beggars are on the corners, because nanotechnology-based systems allow equitable access to shelter, fuel, food and clean water. Electricity and telephone cables have vanished. No wood or coal is being burnt to blacken the sky and promote asthma and tuberculosis. Each house is small but immaculately designed, has a garden, a water tank and a solar fuel, food and fertiliser unit constituting its roof and walls. The air is clean and the number of people seems considerably less. Domestic nano 'assemblers' make whatever household products cannot be created by local communities according to software involving the best, copyrighted international designs. The fact that such visions of technology-reorganised society rarely eventuate as planned does not undermine how valuable they've been as mechanisms of reform.

This is the type of nano-world prophesied by K Eric Drexler in his book *Engines of Creation*. Drexler believed that nano 'assemblers' would soon be bonding molecules together into any stable pattern humans could design (including replicating themselves). As he wrote: 'Nanotechnology will open new choices. Self-replicating systems will be able to provide food, health care, shelter, and other necessities. They will accomplish this without bureaucracies or large factories. Small, self-sufficient communities can reap benefits.' Should a global NES project prioritise assisting to achieve this type of urban renewal outcome?

Such a project appears coherent with most recognised expressions of a global social contract. The United Nations, for example, views access to basic housing as an economic, social and cultural right under the ICESCR. It is also a social right under the constitutional arrangements of many nations. It is emphasised in the *Millennium Development Goals* through the target to provide all citizens with access to basic housing.

In terms of details, a global NES building and manufacturing project might focus on nanotechnology making self-cleaning surfaces for buildings and vehicles. Self-cleaning nano surfaces could be emphasised – drawing inspiration from the lotus leaf that remains clean in a muddy pond. The project might focus on facilitating a wider range of enhancements and applications for such a superhydrophobic surface, for example, using a template that allows nickel nanowires coated with hydrophobic chemicals to self-assemble and trap a layer of air on the surface.

A global NES building project additionally could aim to provide new, less polluting (many anti-corrosion coatings contain chromium and cadmium) approaches to remedying corrosion in building materials. Corrosion inhibitors can be encapsulated, for example within a biodegradable sol gel matrix that provides a protective, passive film/diffusion barrier on the substrate. Building surfaces can be sprayed or painted with coatings containing dispersed nanoparticles chemically treated to acquire a particular attribute, such as dirt resistance, leading to reduction or eradication of the need for toxic solvents. In the UK, researchers at Imperial College London have developed transparent blinds comprising a double-glazed façade, between which are sandwiched lenses that track the sun and focus the light on to nanotechnology-based solar cells with up to 30 per cent efficiency. The system reduces the need for air conditioning, while allowing diffuse sunlight for interior illumination.

Once again, however, a major negative of such a global NES project involves the safety issues associated with nanoparticle use (particularly carbon nanotubes) in building materials. When such structures are damaged, degraded or demolished they could release particles to the air that, when inhaled, might produce chronic lung disease and cancer (carbon nanotubes like asbestos fibres are long and biopersistent and their connection with mesothelioma has been established in animal models). Many, likewise, might consider it a great boon to be able to paint the walls of their house with paint that very effectively resists UV radiation because of embedded zinc and titanium dioxide nanoparticles. Yet significant public health and environmental problems may arise when we attempt to dispose of such clothing, or as such paint begins to flake or needs to be stripped.

A major issue with a global NES manufacturing project would be workplace safety. Fume hoods with exhaust filtration might be installed, protective gloves and eyewear required in protocols for routine use. A specific plastic suit might be worn over normal clothes, antistatic shoes to prevent ignition by static charges and sticky mats at entrances. But the relevant nanomanufacturing companies might (as too often happens now) specify practices in guidelines to protect samples rather than workers with fume hood fans often being turned off when handling nanoparticles. They also might have no nanospecific waste disposal practice and not provide customers with information on the safe use of products containing nano-materials.

A global NES manufacturing project, then, will particularly have to address the issue that nanotechnology does not fit easily into standard occupational health and safety (OH&S) approaches. First, these were usually based on exposure calculations involving calculated mass and mass concentration, whereas nanoparticles often exhibit greater toxicity at lower concentrations owing to their high surface area to mass ratios and enhanced surface reactivity. Second, many structure-activity models used to extrapolate and predict toxicity are outdated. Third, many nanoparticles are being manufactured and used in small start-up businesses whose OH&S standards are difficult to monitor and who lack the resources and expertise to comply. The speed of development of appropriate regulation is also very slow owing to scientific uncertainty about nanotoxicity and reluctance by corporate-funded regulators to implement the precautionary principle.

These problems can be overcome by a global NES manufacturing and building project. Aerosol nanoparticles, for example, have low inertia; yet appear to follow established laws of aerosol physics and classic fluid dynamics. Thus, a well-designed exhaust ventilation system with a high efficiency particulate air filter could remove them from the air. Standard laboratory respirators should be adequate and the evidence appears to be against any 'thermal bounce' leading to increased filter penetration at very small particle sizes. Further studies need to be done on the processing of composites reinforced with engineered nanoparticles that could be subject to kinetic processes such as drilling, grinding, sanding or etching.

Many other issues will complicate public and policy acceptance of a global NES building project. To what extent, for instance, should a global NES project focused on housing or buildings emphasise urban planning? Should its aims involve improving walking and cycling infrastructure; developing and supporting community hubs; reducing use of old photo-synthesis fuel-dependent cars; supplying hybrid or electric cars for fleet vehicles, improving urban design, including street trees, pedestrian cross-ings, more footpaths, allowing reduced distance to public transport and more urban green space?

A global NES building and manufacturing project might offer the building industry production processes that consume fewer resources and cause less biosphere pollution. But that may provide no more than a stop-gap solution to some of the great human and environmental problems we now face. It may be feasible to have a trillion people (in similar bodies as we now inhabit) on Earth, but it is unlikely that they will each be able to have a house and a garden, or the freedom of environmental movement many associate with flourishing. More likely, such vast numbers will live in small virtual reality blocks within a simulated environment. This may in many ways be more pleasing (and solve many of the clothing and housing issues) but would offer a more regimented and less predictable life without exposure to the physical and spiritual challenge of wilderness.

In more immediate terms, as anthropogenic climate change accelerates, a global NES building project could be promoted as assisting people to achieve short-term important survival goals. These might include housing that is adequately proofed from fire, able to store water, safe from home invasion and energy-efficient. If nanotechnology could allow all housing to generate its own power (through solar panels on the walls and roof) that would be a major achievement of the global NES project that facilitated it. In such a case, however, its aims would have to be broader than buildings, manufacture and urban design.

6.   Equitable access to nanomedicines

> Technologies are acquiring properties we associate with living organisms ...
> we are beginning to use technology ... to intervene directly within nature ...
> this disturbs our deep trust ...
> Technology is part of the deeper order of things.
> But our unconscious makes a distinction between technology as enslaving our nature
> versus technology as extending our nature.
> – W Brian Arthur, *The Nature of Technology*

6.1   NANOENHANCEMENT OR NANOPHARMACEUTICALS?

As an Australian academic working in health care regulation, one of my political heroes has always been Prime Minister Ben Chifley. In Canberra his legacy seems to be all around. The Snowy Mountains hydroelectricity scheme, a few hours' drive south of Canberra, was one of his post Second World War nation building projects. Another was the Australian National University where I work, as well as the public-funded Australian Broadcasting Commission (ABC), which over the years has provided balanced TV and radio coverage of many of the major policy debates in which my research has been involved. Just before starting medicine at Newcastle University I lived for a time just a few houses up from Ben Chifley's home in Busby Street, Bathurst. I was married in the Kurrajong Hotel in Canberra, by coincidence (my wife chose the location) the one where Chifley lived while Prime Minister and died while in opposition.

Ben Chifley was what the historian Manning Clark liked to call a 'light on the hill' man. He believed in a world where all people could have life and have it abundantly. Prime Minister Chifley decided after the Second World War to introduce a formulary of free medicines. He did so despite the opposition of the medical profession that funded two constitutional challenges to his legislation, which required a referendum and a change of the constitution to be passed. This medicines scheme became the Pharmaceutical Benefits Scheme (PBS) and its democratic provenance explains why there was such anger that provisions altering its reference pricing mechanism were included at the behest of US pharmaceutical companies in the Australia–US Free Trade Agreement (AUSFTA).

As an academic, my first large competitive research grant involved conducting interviews with members of Australia's Therapeutic Goods Administration (TGA) and Pharmaceutical Benefits Advisory Committee (PBAC) that performs the expert cost-effectiveness assessments for the PBS. The purpose was to examine what if any impact the AUSFTA might have on the capacity of those regulatory processes to respectively provide safe and cost-effective provision of prescription pharmaceuticals in Australia.

Multinational pharmaceutical companies, I discovered through a series of qualitative interviews with senior executives and government officials, use lobbyists and such trade negotiations to further their interests by a variety of techniques that erode the public good. These include increasing the terms of patent monopolies whilst simultaneously claiming long periods of exclusivity for the data whose rapid public dispersal was supposed to be the primary justification for the monopoly privilege in the first place. Other tactics included restricting competition through systems imposed on regulators requiring them to notify patent holders of impending generic products, creating fake medical journals and briefing drug representatives to encourage doctors to prescribe outside safe or cost-effective indications.

I discovered that the experts working at the TGA (Australia's equivalent of the US FDA) were funded by payments from the industry whose safety they were supervising. I found that soon after the AUSFTA lobbying the multinational pharmaceutical industry moved the PBAC in that private sector funding direction as well. Members of the public were surprised when informed that key regulatory agencies in protecting the public interest were reliant on money provided in this way. They could see the potential conflicts of interest that might arise.

This scepticism appeared justified when later under another research grant I studied the processes whereby the TGA declared that nanoparticles in sunscreens were safe. That research revealed how one marketed sunscreen approved by the TGA contained nanoparticles in the uncoated anatase form of titanium dioxide that were causing photocatalytic degradation of steel roofing materials when rubbed off on them from the workers' hands. It was an even more surprising result that when presented with such information the TGA didn't instigate an investigation, make a public recall of that sunscreen product or notify the public of its brand name (though all such actions seemed justified by the precautionary principle that the TGA in its published protocols nominally supported). Recently a committee associated with the TGA went a step further in agreeing (in response to a complaint by a competitor) that a sunscreen manufacturer could not place

'not-nano' in its advertising (although this was factually correct) supposedly because it elicited fears in the public not warranted by the evidence.

All this was quite dismaying to me. One of the reasons that as an
academic I moved into the area of nanotechnology regulation was to get
ahead of the supranational pharmaceutical industry in its attempt to shape
democratic governance systems around the world to its own private advantage. Nanomedicine seemed an area where the options for public interest
oriented regulation were more bountiful, not yet enclosed by private sector
lobbying. Nanomedicine still appears an area strongly worth exploring as
the primary focus of a global NES project.

One-third of the world's population (about 2 billion people) lacks access
to essential medicines – those crucial to preserve life or prevent morbidity
that significantly impedes quality of life. This figure rises to 50 per cent of
the population in certain areas of the developing world. The fact that
intellectual property and international trade laws keep these medicines
away from the patients who need them is fundamentally incoherent with
rational conceptions of science-based natural law and the global social
contract. It also suggests that one mechanism for the governance aspects of
a global NES nanomedicine project is a multilateral treaty in which signatory nations undertake obligations to make the development of nanomedicine coherent with those underlying principles.

Let's use a thought-experiment to explore some ways in which a global
NES nanomedicine project might be framed. Imagine the Emergency
Department (ED) of a Jewish hospital in Iran in the year 2050. The life of
the head of the ED, Dr Abdul, has been a lot easier since the nanotechnology revolution. There is certainly a lot less unanticipated disease and not
just because of enhanced nanocomputer access to information about gene
profiling and gene therapy. The bodies of all his patients now contain
minute sensors that automatically record and transmit biochemistry and
physiological data for analysis in his institution. The nanopharmaceuticals
he and his team prescribe are more like nanolaboratories in that they not
only sense biochemical and physiological anomalies, but also use basic
components in the body to manufacture drugs to treat them. The clothing
of his staff records each doctor–patient encounter for later quality
appraisal. Abdul, though initially sceptical, now thinks of nanotechnology
particularly when applied to health care, as aligned to *Nafs Mutmainnah*
and the striving for ultimate peace.

To achieve public and policy support for a global NES nanomedicine
project it is reasonable to presume that some aspects of such a vision (of
doctors using nanotechnology to more effectively relieve patient suffering)
will have to be widely supported. Much academic and policy discussion

about nanomedicine, however, has emphasised another area – the potential for nanoenhancement of human beings.

Nanoenhancement is usually defined as referring to the capacity of nanotechnology to interact with human structures at the atomic or genetic level to not only prevent or treat disease but also to improve capabilities such as intelligence, athletic ability and appearance. Nanoenhancement, to the extent that it is currently considered by the public or policy-makers, tends to be associated with highly speculative visions of super-humans, such as students inserting a memory stick into their cerebral nanointerface to download textbooks of data directly into their memory. It certainly creates the opportunity to imagine human beings as if from eternity. There is little doubt that the capability of nanotechnology to gradually increase our physical and mental capabilities is worth supporting.

To evaluate whether this provides reason enough to make it the central focus of a global NES project, let's examine an example of nanoenhancement closer to present reality. Cochlear ear devices involve a nanoscale interface between a hearing device and the human auditory nerve. Nanoenhancement can be involved here through the use of nanosilver antimicrobial coatings, nanotubes that release growth factor chemicals (neurotrophins) and pattern nervous stimulation, as well as nanofibre stimulating electrodes. Another example involves superparamagnetic iron oxide nanoparticles with magnetite (Fe_3O_4) that could transport hearing-enhancing molecules to the inner ear. Collectively such nanotechnology could not just repair damaged hearing organs, but could also improve the efficiency of the auditory nerve for the whole human population.

If a global NES project does focus on such forms of nanoenhancement, however, it is likely to confront strong moral objections concerning whether it is prioritising autonomy (freedom of choice) over issues of social equity (the opportunity cost of diverting public funds to this area). There is, for example, no potentially catastrophic public health or environmental disaster to which nanoenhancement is directly addressed. Nanoenhancement does not attempt to resolve the number of people dying of poverty, the other species becoming extinct or ecosystems being permanently degraded. When seen from the perspective of the ethical and legal foundations of a global social contract, in terms of how human beings should operate from a universal or eternal perspective, nanoenhancement quite justifiably may appear a largely self-indulgent exercise at this time and one insufficiently coherent with foundational social virtues such as justice, equity and environmental sustainability.

For such reasons nanomedicine (focused on the primary professional virtue of loyalty to the relief of patient suffering) is a potentially more successful candidate for a global NES project. It is uncontroversial to state that nanotechnology already is an expanding area of pharmaceutical research and development globally. More than 200 drug companies have active research programs in nanomedicine. Such initiatives are mostly predicated on nanotechnology having a powerful enabling function that will enhance the efficacy and market competitiveness of existing pharmaceutical and medical device products.

Nanotechnology offers particular value as an improved drug carrier and delivery system able to get precise concentrations of drug to selectively targeted effect sites in the body. It might achieve reduced toxicity by requiring that much smaller doses of existing drugs be used. It could facilitate production of systems for slow and continuous release of drugs where that is necessary. It may support the building of scaffolds that support tissue growth. It could allow enhanced imaging for detection and treatment of disease. Peptide nanotubes, for instance, have been investigated as the next generation of antibiotics, and as immune modulators. Anti-cancer drugs are another particularly strong field for nanomedicine research. Abraxane™ (paclitaxel albumin-bound particles), a nanotechnology-based anti-neoplastic agent, provides an illustrative example. Abraxane, as with many nanotherapeutics, constitutes a nano-reformulation of a pre-existing medicine. Its active ingredient paclitaxel is an anti-microtubule chemotherapeutic agent from the taxane group.

Nanosilver now is used extensively in hospital settings for wound management, particularly for the treatment of burns and ulcers. Nanosilver particles in the 1–10 nm range have been shown to inhibit binding by HIV-1 to host cells. Nanosilver is utilised to coat urethral and central line catheters as well as other implantable medical devices to prevent the growth of slime-containing biofilms that promote bacterial infection and sepsis.

In terms of governance requirements, increasing numbers of drugs involving nanotechnology are being approved by the United States Food and Drug Administration (FDA) and other drug regulators globally for human use. A significant regulatory issue here is that nanodrugs and nanomedical devices blur regulatory distinctions between drugs, devices, biologics and combination products with a convergence of electronics, materials and biologics science unsettling traditional physico-chemical rules.

There are three other areas where nanomedicines may create unusual challenges for existing pharmaceutical industry governance systems and structures. The first and perhaps most obvious concerns the requirement for the suppliers of a substantially more costly nanomedicine to establish its substantial reduction in toxicity over comparitors. The second relates to comparisons of a nanomedicine on effectiveness and cost against existing marketed therapies. The third point concerns the heightened potential for anti-competitive behaviour associated with the unusual capacity of lobby-ists for nanomedicine manufacturers to make claims to 'innovation'.

The current literature on nanotoxicology strongly suggests a level of agreement about adverse effects of small size, surface area, insolubility and kinetic effects of engineered nanoparticle (ENP) damage at the cellular level. Some ENPs have been shown to preferentially accumulate in mito-chondria and inhibit function; others may become unstable in biological settings and release elemental metals.

The US FDA appears to have assumed that macro-scale safety translates to that at the nano level. A nanoparticulate reformulation of an existing drug, for example, has been deemed by the FDA not to require an Abbrevi-ated New Drug Application (ANDA) because bioequivalence was deemed already established by the research required for regulatory approval of the parent compound. Thus Abraxane, the nanoreformulation of paclitaxel mentioned earlier, could claim a reduction of toxicity because its form as an injectable suspension evades the hypersensitivity reaction associated with Cremophor EL, the solvent used in the original macro compound. A major challenge for regulators here is that nanomedicine toxicological effects may not readily be predicted by extrapolation from macro-scale equivalents.

Another challenge is that most contemporary nanotoxicological studies of nanomedicines focus on the interaction with biological systems of the surface of ENPs, not on the fluxing corona of proteins that aggregates around them in human plasma. How this ever-shifting protein corona associates with ENPs is a critically important but largely unexplored factor in how they enter and leave cells and hence their toxicological and thera-peutic fate. The coating of proteins over an ENP in different organ compartments or cellular environments, for example, may transmit altered biological effects owing to altered protein conformation, exposure to novel proximate amino acid residues or epitopes, perturbed function and down-stream cellular signalling pathways (owing to structural effects or local high concentration) as well as avidity effects from close spatial repetition of the same protein.

One governance mechanism behind a global NES project in this area could involve an NES treaty on the research, safety and cost-effectiveness of nanomedicines. Its provisions might create obligations for uniform domestic legislation by ratifying nation states requiring mandatory labeling and separate safety testing for nanomedicines and their bulk equivalents. It could recognise that nano versions of existing chemicals used in nanomedi-cines should be automatically assessed as new chemicals and properly labelled.

An NES nanomedicines treaty, in the face of uncertainty about the toxicological data associated with nanomedicines, could include provisions applying a variation of the precautionary principle. The precautionary principle's core idea, as already mentioned, is that lack of scientific certainty about the potential harm of a product or process being evaluated by regulators should not be used as an excuse to delay measures protecting the public or the environment from harm and should allow marketing of such a product to be appropriately delayed, restricted or have its risks notified to the public.

The extent to which the precautionary principle might apply to decisions by regulatory agencies coordinated by a nanomedicine treaty would depend on the correlation between the nature and seriousness of the risk as well as the kinds of remedy to be made available. For example, if the potential toxicological risk from a nanomedicine is determined (by the treaty's coordinated international assessment process) to involve serious and irreversible threats to human bodies and environment, strong regulatory measures (such as a ban or moratorium on marketing) could be textually justified through each nation's regulatory processes even in the face of scientific uncertainty. If the potential toxicological risk from a nanomedicine is found to be not so serious as to prevent marketing approval, it would still support a precautionary measure requiring ongoing post-marketing toxicological investigations with reporting requirements.

One of the main governance challenges for nanomedicine that must be addressed under a global NES project in this area relates to the fact that it is unlikely manufacturers of such medicines will provide data showing cost-effectiveness over a long time frame in terms of QALYs (Quality-Adjusted Life Years), or in a head-to-head randomised clinical trial (RCT) against an existing marketed product for the same indication. A global treaty on safety and cost-effectiveness of nanomedicines could require that pharmaceutical companies must provide all clinical trials relevant to the listing, not merely those considered 'pivotal'. Failure to disclose all studies would create a fatal flaw for a regulatory safety submission. Such a treaty could go further and include not just a requirement that RCTs be disclosed, but specify the type of RCTs that should be conducted. Similarly, the treaty (or guidelines prepared under the processes it establishes) could provide that, in circumstances where there is no comparator treatment available, the 'current standard care' is to be the objective reference point.

A related challenge will be whether manufacturers of nanoreformulations of listed drugs should be required to provide head-to-head RCTs comparing effectiveness against the original compound rather than a placebo. Manufacturers of nanomedicines, protective of their investment in this new field, will be reluctant to endorse the former option. The prices for nanomedicines are likely to be particularly high in order to recoup actual and hypothesised research and development costs. It also may be unusually difficult for companies manufacturing nanomedicines to enrol human subjects into the various types of clinical trial required to constitute the necessary evidence. The incompletely understood risk profile of nano-medicines (a problem that is likely to remain unsolved for some time) may cause patients enrolled in research trials to be reluctant to remain on nanomedications long term and there may be a heightened tendency for nanomedicine RCTs to use more rapidly acquired data such as physio-logical parameters (for example reductions in biomarkers) as outcome measures.

An NES nanomedicine treaty also might require that the procedures of regulatory agencies specify that where an alternative pharmaceutical is 'substantially more costly' scientific experts shall not recommend it for government subsidy unless 'for some patients' it represents a 'substantial improvement in efficacy or reduction in toxicity' over those alternatives. This could be vital in reassuring the public of the value of such a project given how highly reactive and mobile ENPs may present unique health risks when used in medical applications.

Each step of a globally coordinated nanomedicine safety and cost-effectivness process should remain as rigorously objective as possible and close to the well-recognised standard of biological equivalence. The NES nanomedicines treaty could require that all citizens of the world have access to safe and efficacious medicines at 'a price affordable to the individual and community' and that process of scientific evaluation should seek to fulfil these objectives through assessing a new drug's quality, safety and efficacy (QSE) along with its clinical cost-effectiveness.

An interesting case study of how nanomedicines could be evaluated for innovation under an NES nanomedicines treaty is provided by Annex 2C of the Australia–US Free Trade Agreement (AUSFTA). Australia's interpret-ation of the constructive ambiguity of 'innovation' defined the concept more as 'health innovation' (the national benefit proven by expert analysis of scientific publications to arise from marketing the product) rather than the US interpretation of pharmaceutical innovation as requiring 'market-ing innovation' (the degree to which lobbying and advertising can make a case for its technological innovation).

How regulatory systems deal with claims that nanomedicines deserve subsidies because they are innovative will be important to their global regulation. It is expected, for example, that nanoreformulations of older drugs will appear on the market earlier than novel 'blockbuster' nano-based drugs. Such 'me-too' drugs may offer only a minor (if any) therapeu-tic benefit despite their novel nanotechnological base. The treaty could define pharmaceutical 'innovation' (as it was in Annex 2C of the AUSFTA) on the basis of objectively demonstrated therapeutic significance assessed through the operation of science-based cost-effectiveness systems, as well as by the operation of competitive markets facilitated by strong anti-monopoly laws.

An NES nanomedicines treaty could include 'anti-evergreening' provisions and allow the manufacture of generic medicines for export to international markets where relevant patents have expired. It could expand the compulsory licensing exceptions that allow nanodrug patents to be broken (with reasonable compensation) by domestic generic manufacturers in a public health emergency, allow 'springboarding' by generic nanomedicine companies on patent expiry, and permit the research use exemption that allows public-funded university researchers to experiment with the chemistry of drugs that are in patent without having to pay royalties.

Yet nanomedicines, however effective they may be, are modes of disease treatment. A global NES project focused on nanomedicines (for example through their research, development, safety and cost-effectiveness regulation) would leave relatively untouched other major causes of disease and mortality such as lack of food, water and energy for cooking and heating.

6.2 INHIBITING CORPORATE FRAUD UNDER A GLOBAL NES HEALTH CARE PROJECT?

I enjoy my academic and policy contests against Medicines Australia (MA), the lobby group for the patented pharmaceutical industry in Australia. MA is a politically powerful organisation. The legislation it successfully lobbied for to split the Australian prescription medicine cost-effectiveness formulary into patented and generic medicines groups was rushed through both houses of parliament in a few weeks after a Senate inquiry that lasted a few days. They even got the federal government to sign a Memorandum of Understanding with them to cut the price of drugs in their competitors' (the generic medicines) sector. After the latter policy triumph, I received my first Christmas card from Medicines Australia. It seemed time, in the public interest, to try a different tack to counter their private interest lobbying. The new approach I took is likely to have benefits to the governance arrangements of any global NES project, but particularly one involving health care or medicines.

The particular research grant I decided to apply for (and was subsequently funded to undertake by the Australian Research Council) involved investigating how a form of the US *False Claims Act* could be introduced into Australia. The *False Claims Act* involves a mechanism for rewarding acts of conscience that provide insider information that is critical for successful anti-fraud actions by the government against the corporate sector.

The *False Claims Act* was initiated by President Lincoln during the US Civil War as a means of preventing the government from being defrauded by military contractors. It draws upon an ancient common law doctrine that allows an individual citizen to bring a legal action on behalf of the public good (such as returning defrauded public monies). It then provides for financial rewards and/or compensation to that person if the action is successful. The *False Claims Act* has been very successful. Each year billions of dollars are recovered, in particular from pharmaceutical and medical device companies, after successful anti-fraud claims under this legislation.

Health care and medicines, because of the substantial private sector profits that are made, are the largest area of *False Claims Act* practice in the US. One important governance aspect of a global NES project focused on health care or medicines could involve a multilateral treaty with model provisions (for incorporation in domestic legislation) encouraging informants to reveal corporate fraud in relation to the large amounts of taxpayer money that may be spent under such a project. WTO agreements would allow such coordination of governance approaches against corporate fraud. This goes further than merely suggesting that informants about fraud under a global NES project should be protected from unjust reprisals after their public interest disclosures, that NES project research participants shouldn't be forced into waivers of rights or releases from liability, or that NES researchers must disclose financial conflicts of interest or scientific misconduct.

The major advance against corporate fraud that such an approach could promote in a global NES project (particularly one on health care and medicines) involves a version of the *Qui Tam* provisions of the *False Claims Act*. The term *Qui Tam* is an abbreviation of a Latin legal maxim broadly meaning 'he who sues on behalf of himself also sues on behalf of the state'. *Qui Tam* laws allow private citizens (called 'relators' under the legislation) the right to disclose 'corporate insider' documents establishing fraud to a 'no win, no fee' lawyer and initiate a lawsuit under seal (not initially disclosed to the defendant).

The informant is potentially rewarded with 15 to 30 per cent of whatever proceeds the government recovers from the civil suit. The prospects of success are greater, however, if federal or state justice department officials can be convinced to join the case. In such instances the *Qui Tam* relator and his/her counsel may contribute valuable manpower and financial resources to the action. With the FCA providing for treble damages since the 1986 amendments (held by the US Supreme Court to be compensatory and not punitive in nature), the legislation provides a substantial financial incentive to get crucial documentary evidence about fraud out from inside the corporate sector that overcomes concerns about the usual risks associated with whistleblowing (such as intimidation, loss of livelihood, friends, family and mental disability).

The range of fraudulent activities in health care covered by the FCA is broad and extends to prohibited conduct prescribed in other federal statutes, such as the *Anti-Kickback Statute 1972*, the *Stark Law Statute 1995*, the *Patient Protection and Affordable Care Act 2010* (PPACA) and the *Food, Drug, and Cosmetic Act 1938*. In the words of the United States Supreme Court, 'the [FCA] was intended to reach all types of fraud, without any qualification, that might result in financial loss to the Government'. The enforcement partnership has proven extremely cost-effective, recouping $15 for every $1 spent on *Qui Tam* investigations and litigation. *Qui Tam* laws can act as potent deterrents for fraudulent activities as they amplify the threat of detection and prosecution and create incentives for compliance with governance standards. The relevant government law enforcement body retains control.

Where the government decides to intervene in the action, it takes over the prosecution of the claim and the relator must tender its full cooperation or the government may compel the court to limit the relator's role in the litigation. Even where the government refuses to intervene so as to allow the relator to proceed with the lawsuit on the government's behalf, the government gets to actively monitor the case, has a right to review all pleadings and to later join the case where 'good cause' is shown. As such, fears associated with the 'privatisation' of public enforcement, including those about perverse incentives driving enforcement or over-enforcement, appear unfounded.

Frivolous or 'parasitic' *Qui Tam* claims are prevented by statutory bars to individuals who have made no material contribution to uncovering the fraud or providing the factual basis of the case. Secondly, a relator cannot base his or her *Qui Tam* action on publicly disclosed allegations, unless the relator is the 'original source' of that information. The PPACA has broadened the ambit of 'original source' to include those who merely have an indirect knowledge of the fraud, provided the relator has 'independent knowledge that materially adds to the publicly disclosed allegations'.

The model has been so successful that nearly half of all US states have enacted their own false claims statutes. It has, however, yet to be picked up internationally and doing so could be a major advance in NES project governance arrangements regardless of whether nanomedicine is the major focus of the project.

A global *False Claims Act*-type anti-corporate fraud mechanism might become an important part of any global NES project: for ensuring the substantial public monies dedicated to it are not misused. It would be a valuable subsidiary governance mechanism encouraging transparency and accountability, but not necessarily the central forward vision of how major global public health and environmental problems can be solved using nanotechnology.

6.3 MARRIAGE TO PUBLIC OR ENVIRONMENTAL GOODS BY CORPORATIONS IN NANOMEDICINE?

Along with Michael Moore's *Sicko* (about the denial of health care to US citizens *who have* private health insurance), one of the most fascinating documentaries I've watched was called *The Corporation*. The basic premise of the film was that corporations have been accorded status as 'people' under the law of most nations. This allows them to sue and be sued and even to bring actions for defamation in some jurisdictions. Yet, so the film argues, if a psychological personality profile was performed upon a corporation it would probably come to the conclusion that the subject was an extremely narcissistic individual. One of the major reasons for this, the film explains, is that corporations are legally required by their governance arrangements to maximise shareholder profit as their dominant concern. Imagine if a real human being lived or behaved like this. He or she would be a caricature of the worst type of greedy financial speculator, single, manipulative, always focused on the bottom line, and ambitious to the point of ruthlessness.

I have argued in previous chapters that the WTO as well as trade and investment treaties will create major obstacles for a global NES project because they are allowing global governance responses to the most important public health and environmental challenges humanity has faced to be organised by these narcissistic, artificial, corporate persons. It is now time to look at some radical remedies for this problem.

As we've mentioned, there are some important reasons why such a narcissistic personality profile, at least in governance theory, doesn't apply to sovereign nation states. The main one is that the governance of such states is predicated on a social contract involving constitutional arrangements (purportedly modelled on ideal symmetries in nature and human relationships according to science-based natural law jurisprudence) that charge the government with respecting the rights of individuals whilst it strives to achieve goals such as ensuring the right to health, education, peace and security. Yet, the more military or corporate cliques take over the control of states and their finances, and the more that within such geographical borders people exist in slums without rights or services, those entities can rightly be considered to be failing. It follows that in an area as important as health care and medicine a global NES project should directly address these issues and find a way to transform the ethos and primary motivation of the corporate actors likely to be involved.

Let's provide some background as to why such a governance reform could be so important to the success of a global NES project. The enlightenment philosopher Immanuel Kant concluded that if we stop to think of all the care that afflicts us in our search for ways of passing life as short as this, and of all the injustice that is done in the hope of a future enjoyment that will last for so short a time, a life expectancy of 800 years or more would *not* be to our advantage. He claimed that inheritance would be delayed and fathers would live in mortal fear of their sons, brothers of brothers and friends of friends, and the vices of a human race with such longevity would necessarily reach such a pitch that it would deserve no better a fate than to be wiped from the face of the Earth. There can be little doubt, however, that longevity will be a major focus of corporate efforts in nanomedicine, if only because it caters to the need of a presumptively affluent population whose payments will increase the profits of corporate entrepreneurs and the entities they work for.

Large-scale corporate investment will be vital to the success of a global NES project. Yet, the corporation is by its legal constitution required to be a profit-seeking entity. This means under current legal structures that it is inherently greedy. The responsibilities of corporations are not set out in national constitutions as they would be if their role had been the subject of social contract negotiations. As previously detailed, the comparatively unrestrained linkage of the interests of supranational corporations with global trade and investment has severely impacted on the capacity of individuals and their communities to shape the course of world events towards environmental sustainability as well as increased human dignity and survivability.

One approach that a global NES project could adopt to remedy this issue might involve exploring methods by which supranational corporations could be 'married', in the sense of being bound ethically and legally to prioritise the interests of others in ways acceptable to national and global social contracts. A practical vehicle for facilitating corporate 'marriage' could be a stronger public interest requirement of corporate registration specified by a United Nations treaty.

Such international obligations, for example, could be part of a treaty establishing a United Nations World Environment Organization (WEO) (particularly for its benefits to global public health) that would unify and enhance the efficiency of existing piecemeal global environment regimes so they could compete equally with institutional pro-corporate behemoths such as the WTO. This might happen through the concept of linking environment protection funding and programs under such a WEO treaty to a tax on global financial transactions.

After the law reform issue of 'same sex marriage' has been resolved (as it must if the level of sympathy and symmetry characterising our social arrangements is to evolve), the next big challenge could be 'corporate marriage'. Global NES project 'corporate marriage' provisions (or those of a treaty such as that establishing a putative UN WEO) could require the constitutive documents of corporations to impose as a legal condition of registration that corporations be in effect 'married' to some acknowledged community or environmental good.

Another example might require involvement by those corporations participating in a global NES health care or medicines project in a Medicines Patent Pool. By this process nanopharmaceutical patent holders would share their intellectual property with the Pool, and then license it to other manufacturers to facilitate affordable generic medicines in resource-poor settings.

Such innovative governance mechanisms offer the prospect of transforming a corporate artificial person. Previously created and driven by corporations law into the social isolation implicit in absolute prioritisation of shareholder profit (thus becoming like some creation of Mary Shelley's Victor Frankenstein) corporations could mature into global citizens with firm legal responsibilities creating a geometric network of relationships that transmute self-interest. The rate-limiting step in economic growth might then become simply the power of corporate executives to imagine how the new technologies their 'married' companies are developing will work to create a sustainable world that promotes the flourishing of life.

Likewise, the consensual nature of the place of corporate-controlled intellectual property (patents in particular) under a global social contract could be emphasised in the governance arrangements of a global NES health care project as well as a UN WEO treaty. The balancing of interests involved is probably better served by referring to patents as a form of 'intellectual monopoly privilege' (IMP). Such terminology recognises that innovative ideas do deserve to reap profits through a brief legally sanctioned period of market monopoly, but that it is not a form of 'natural' right that accrues to the corporate artificial persons who purchase them; rather, it is a privilege they are granted insofar as society also benefits from the consequent rapid distribution of knowledge. This reconceptualisation of the public role of patents might become an important part of the global governance framework into which nanotechnology innovations must fit if they are to become a major contributor to a sustainable world.

The legally required marriage of a supranational corporation could be to a community, or to an aspect of a global public good. The corporate registration process would involve a contract as to what were the appropriate terms, monitoring and outcome measures of such a marriage. The capacity for 'divorce' and 'remarriage' of artificial corporate persons could, of course, also be specified. Once again, however, such a governance approach, while likely to be a beneficial adjunct of most global NES projects, will not of itself provide the motive force to tackle major public health and environmental problems.

Counterarguments would be that virtue (as we demonstrated in earlier chapters) is acquired by voluntary or free willed decisions to apply universally applicable principles in the face of obstacles. This suggests that the creation of social and financial incentives might be better than legal compulsion as a means of forcing supranational corporations into practical engagement with global public goods. Such thinking, it is argued, however, involves applying human considerations to artificial persons and a logical extension of it may get us back to the same place – that some form of arranged relationship between supranational corporations and global public goods is necessary and coherent with the revitalised basic form of a global social contract.

7. Nanotechnology for global peace and security

> The truth is, that no mind is much employed upon the present:
> recollection and anticipation fill up almost all our moments ...
> Of the uncertainties of our present state,
> the most dreadful and alarming is the uncertain continuance of reason.
> – Samuel Johnson, *Rasselas*

7.1 RESEARCHING MILITARY NANOTECHNOLOGY FOR A SUSTAINABLE PEACE?

My family, on my father's side, has a long tradition of military service. Direct relatives fought against the French at Quebec, against Napoleon in Spain and with the Australian Light Horse in the battles of Gallipoli, Beersheba and Romani in the Middle East during the First World War. I've always thought that reading military history gives you very valuable vicarious experience, especially for an academic career. Such history also shows how military needs and expenditure have been responsible for major advances in human technology. Peace and security undoubtedly are major public and environment goods coherent with a global social contract.

The philosopher Immanuel Kant (whom readers by now will have discerned is one of my favourites) stated that the greatest evils that oppress civilised nations are the result of war – not so much actual wars in the past or present as the unremitting, indeed ever-increasing *preparation* for war in the future. Most of the resources of a nation state and the fruits of its culture and industry (which might have enhanced the well-being of its peoples) continue to be devoted to this military purpose. As a result, the freedom, happiness and prosperity of citizens (though ostensibly thus protected from real and immediate or concocted threats) also suffer greatly as the state's care for its vulnerable members is subjugated to the demands of companies making military products and the government departments they lobby.

Governmental secrecy surrounding research and development of military technology makes it difficult to describe the current level of military applications of nanotechnology with any degree of certainty. There can be no doubt, however, that nanotechnology is a major focus of strategic interest and will revolutionise any offensive and defensive military capabilities – it is appropriate, in other words, to consider the viability of a macroscience NES military nanotechnology project.

The military use of nanotechnology is already a significant global phenomenon, as illustrated by the funding poured into military research and development in nanotechnology in the US, the UK, India, Sweden, Russia and China. In 2001, for example, the US established the National Nanotechnology Institute (NNI) as an inter-agency program to coordinate federal research and development activities in nanotechnology. The NNI allocated US\$460–464 million in fiscal year 2008–2009 and proposed US\$379 million for fiscal year 2010 as investment in nanotechnology research and development in the Department of Defense. The UK initiated its military nanotechnology program on a much smaller scale, investing £1.5 million in 2001. Sweden as another example has reportedly invested €11 million over five years in nanotechnology research for military purposes.

More recently, India has sanctioned at a total cost of Rs12.48 crore under the Armament Research Board in nanotechnology-related fields such as high energy materials, armament sensors and electronics, ballistics, aerodynamics and detection of explosives. India's Defence Research and Development Organisation plans to establish five centres of excellence, including a centre for nanotechnology-based sensors for WMD detection, and a centre for nano optoelectronic devices, each budgeted at Rs50 crore over five years. Although figures are not made public, Russia has also reportedly been significantly investing in nanotechnology that will enable new offensive and defensive weapons systems.

The private sector also has considerable investment in military nanotechnology. The Institute for Soldier Nanotechnologies (ISN) is a research collaboration between the United States Army and the Massachusetts Institute of Technology (MIT), combining basic and applied research into military applications of nanoscience and nanotechnology in protection, performance improvement, injury intervention and cure. Private companies such as QinetiQ, BAE Systems, Lockheed Martin, GE, Industrial Nanotech Inc., and Raytheon are involved in the research and development of military nanotechnology in partnership with the government, especially in the areas of nano-sensors and body armour, electronics, optoelectronics, and information and communication systems for detecting, preventing and deterring bioterrorism. An advanced armour-piercing projectile involving the potential use of NanoSteel™ was recently patented in the US.

A global NES project in this area would likely concentrate on detection of strategic risks and threats. Nanotechnology sensors and computing, for instance, are being designed to assist in uncovering bioterrorist threats of aerosol attacks on individuals or crowds, 'dirty' bombs and targeted contamination of food sources. Bioterrorist attacks involving contamination of milk with botulinum, or release of pathogenic organisms and biotoxins in the water supply, might not themselves involve nanoscale agents. Their detection, however, may require correlation of vast amounts of information beyond the capacity of non-nanotechnology sensing, information and communication systems. Likewise, threat responses to unexpectedly virulent modifications such as mousepox IL-4, or a highly virulent strain of the influenza virus (akin to the strain that caused the Spanish influenza pandemic in the winter of 1918–1919 and killed up to 50 million people worldwide), are likely to greatly benefit from defensive nanotechnology surveillance systems.

Atlantic Storm was a simulated bioterrorism exercise based on the deliberate release of smallpox viruses in various European and North American cities. It revealed that many nations had inadequate vaccine stockpiles, response plans and public health laws to effectively respond. Such exercises emphasise the need to develop innovative defensive nanotechnologies capable of allowing health workers to promptly detect minute viral loads in widely dispersed locations and effectively communicate the relevant details to public health authorities.

Nanotechnology also has defensive military applications in relation to lighter, stronger and more heat-resistant armour and clothing, bio/chemical sensors, lighter and more durable vehicles and miniaturisation of communication devices. The US *Defense Advanced Projects Agency*, for example, funds projects in nanotechnology-related magnetic memory, biocomputing, bio-molecular motors, sensors for chemical and biological warfare agents and microrobots. Carbon nanotubes have been used as gas sensors. Soldiers capable of detecting biological hazards, chemicals or even the presence of other combatants through nanosensors in clothing with strength enhanced by carbon nanotubes will have considerable combat advantage. Nano-uniforms will also be able to allow commanders to detect the position of soldiers and medics to diagnose and treat wounds from a distance.

Major components of a global NES military project are likely to involve producing weapons. Nanotechnology allows the building of conventional missiles with reduced mass and enhanced speed, small metal-less weapons made of nanofibre composites, small missiles and artillery shells with enhanced accuracy guided by inertial navigation systems and armour-piercing projectiles with increased penetration capability. Although they are still highly speculative, a global NES military project could lead to the development of micro-combat robots, micro-fusion nuclear weapons, new chemical agents carried by nanoparticles, and new biological agents with self-replication capability.

Carbon nanotubes may be developed under such a project as lightweight, heat-resistant coating for intercontinental ballistic missiles. In explosives, nano-sized oxidiser and fuel materials with a high surface area offer the potential for rapid energy release. Combining nanometals such as nanoaluminium with metal oxides such as iron oxide produces faster chemical reactions, with greater explosive energy released more rapidly.

Some of the potential, offensive military applications of nanotechnology under a global NES military project could span several traditional technological compartments and blur the distinction between conventional weapons and weapons of mass destruction. The ability of nanotechnology to design and manipulate molecules with specific properties could lead to biochemicals capable of altering metabolic pathways and causing defined hostile results ranging from temporary incapacitation to death, or controllable, multilayered biochemical weapons that might limit and regulate the spread of biochemical agents. Nanotechnology could also make it possible to contain and carry a minute amount of pure-fusion fuel safely until released to detonate a micro-nuclear bomb.

Global governance issues are critical for the appropriate use of military nanotechnology. States negotiating under the United Nations *Biological and Toxin Weapons Convention* (BTWC) recently emphasised the need for broad-based codes of conduct for both scientists and public health physicians to counter future bioterrorist threats, partly by warning of the professional perils involved in deliberate or inadvertent release of information and substances.

Military use of such nano-weapons may be governed by prohibition and inspection mechanisms under existing arms control treaties, as long as currently available chemicals and biological agents are used in nano-size. However, the dual-use potential of nanotechnology and the low visibility of nanoparticles in weapons may make it hard to detect their development and use as weapons. Further, the definition, effects and impacts of nano-weapons are yet to be comprehensively detailed under any of the existing international legal regimes on weaponry. This creates both problems and opportunities for governance developments under a global NES military nanotechnology project. Most weapons manufacturers, for example, employ teams of legal experts designated with ensuring their products comply with existing international arms control standards.

Technological developments with novel military applications likely to arise from a global NES military project inevitably will pose challenges to effective international regulation, not least because of the inevitable secrecy during their research and production. International arms control regimes can be viewed, according to the scheme presented here, as direct commitments to peace and security as foundational social virtues under a global social contract. They have been set up to regulate the manufacture, deployment, use and monitoring of certain types of weapon with major focus on chemical, biological and nuclear weapons. Recently, however, the application of computing and software innovations to various emerging technologies has led to major changes in the military tactics of developed nations, which may have outpaced existing arms control regimes under international law.

The governance aspects of a global NES project focused on military capabilities could create processes and organisational structures to investigate the potentially devastating effects of such nano-weapons. At an individual level, explosives such as those using nano-energetic particles, nanoaluminum or non-metal nanofibre composites, for localised high energy explosions (such as the Israeli armed forces are believed to have used in their 2009 Gaza incursion) could cause unnecessary suffering to both combatants and non-combatants with prolonged health effects. At a larger, strategic level, the development and deployment of smaller, longer range missiles with greater precision, or new biochemical agents could change dramatically the military power balance and the way in which a war is fought.

But it is unlikely that the public will embrace a global NES project focused on military nanotechnology as a policy and funding priority ahead of issues like health care, energy and environmental protection. Whilst peace and security, for example, may be valuable outcomes of such nanotechnology research, so too may there be greater capacity for terror and abuse of power.

7.2   UNITED NATIONS CONVENTION REGULATING MILITARY NANOTECHNOLOGY?

At law school I always enjoyed and did well at international law, in part because I was imaginatively captured by its endeavour to shape the conduct of nations within rules that seemed to emerge from very idealistic thinking about the role and future of humanity. I became passionately interested, for instance, in the way international law scholars such as Richard Falk and officials like United Nations Secretary General Dag Hammarskjold were prepared to envision global governance systems aimed at the betterment of mankind.

I read Hammarskjold's *Markings* as a paean to conscience in the service of world order. Hammarskjold's reliance on the works of medieval mystics including Jan van Ruysbroeck offered an important message of hope and humility. Hammarskjold seems to have interpreted van Ruysbroeck as requiring a big public sacrifice; indeed his probable assassination by state and corporate interests opposed to his United Nations intervention to establish democratic rule in the Congo. I viewed van Ruysbroeck, however, as encouraging meditative absorption in the 'boundless clearness'. Much later (by another of those pregnant coincidences that suggest our dual existence as a particle and wave), after giving a paper on nanotechnology regulation in Leuvan, I was able to visit the site of van Ruysbroeck's old monastery at Groenendaal, in the Sonian Forest.

The efforts of United Nations weapons inspectors operating under international conventions have been a major factor in reducing the risk of catastrophic nuclear war or the use of chemical or biological agents of mass destruction. Imagine a future where invisible nanosensors and nanorobots, developed for the military, have been taken over by law enforcement officers. Nanotechnology-based autonomous killer robots or implants designed to facilitate use of lethal force would call into question many of the basic ethical assumptions underpinning international law's regulation of military action, for example, that a human should be the decision-maker whenever a person might be killed in battle. Such nano-weapons also would make usual verifications systems of many United Nations weapons control treaties obsolete. Nano-weapons production, not needing the usual bulk scale facilities, may be very difficult to locate. A reasonable conclusion is that nanotechnology could have unfortunate potential to produce weapons of mass destruction.

These concerns suggest that a core governance aspect of a global NES project focused on military use of nanotechnology might be a specific international nano arms control convention. Because of the manifest strategic advantages of nanotechnology, it is unlikely such a convention would involve a moratorium or ban. More likely it would be designed to establish a preventative arms control regime based on prospective scientific, technical and military operational analysis of nanotechnology. Such a treaty could address the right balance between military necessity, humanitarian considerations and peaceful applications of nanotechnology as well as processes to ascertain and monitor the extent to which nano-weapons might already be, or can be, researched, prohibited or regulated.

Currently there is no international treaty that has specific provisions regulating nano-weapons. To justify a global NES nano arms control treaty it will be necessary to determine to what extent nano-weapons are covered by existing international law including general principles governing weaponry. States have agreed in a variety of international treaties to specific and express rules relating to arms control, which apply even in peacetime. Yet, the adoption of treaties to prohibit certain weapons tends to be reactive to claims of new effectiveness, rather than pre-emptive as would need to be the case in relation to military use of nanotechnology. Thus, states have agreed to ban the employment of projectiles of a weight below 400 grams that are explosive or charged with fulminating or inflammable substances, expanding bullets, asphyxiating, poisonous or other gases, biological weapons, chemical weapons, blinding laser weapons, anti-personnel mines, and most recently cluster munitions.

Nano-weapons developed under a global NES military project, as mentioned, could be regulated at least in part by the relevant existing arms control conventions. For example, prototype nanolasers producing megawatts of continuous power are far more powerful than those previously known, which are likely to be subject to the 1995 *Protocol on Blinding Laser Weapons* in the visible region. Nanotechnology can also produce toxic chemicals with novel properties, and may facilitate the development of synthetic organisms with a high degree of lethality. Yet these treaties were drafted without any consideration of nanotechnological developments and therefore may not be sensitive enough to the enhanced and novel ways in which nanotechnology may assist arms development even in these areas.

The recent deployment of Dense Inert Metal Explosives (DIME), for example, illustrates the difficulty in defining whether new weapons (developed for example under a global NES military project) would fall within the nanotechnology category, or within existing rules of arms control law. DIME was engineered at the US Air Force Research Laboratory from depleted uranium research in order to achieve low collateral damage by producing a highly powerful blast within a relatively small area. It is the latest innovation in the US military's long-running development of Focused Lethality Munitions (FLM), designed to provide the 'weapons of choice' in targeting terrorists hiding among civilians.

Upon detonation, the DIME carbon-fibre warhead case disintegrates into minute, non-lethal fibres with little or no metallic fragments, then spraying a superheated micro-shrapnel of powdered (potentially nanoscale in effect) tungsten particles with sufficient penetration mass for disabling the target within a small lethal footprint. This weapon takes a step further into the future in which metal dust can be blasted directly into a human body, rather than spraying or exploding clouds of fragmented shrapnel. It may reduce collateral damage but also complicate and negate surgical treatment of resultant wounds in contravention of humanitarian law treaties. DIME bombs were reportedly employed by the Israeli military during the 2006 conflicts in Gaza and southern Lebanon, and more recently during the Gaza conflict in January 2009. Israel is a party to the 1980 Protocol I.

Owing to the undetectable nature of tungsten micro-particles in human tissue, the question arises whether a nanotechnology weapon similar to this would fall within the scope of the 1980 Protocol I to the *Convention on Conventional Weapons* (CCW) on Non-Detectable Fragments. To do so the design intent must meet the threshold for the prohibition, for example, because the primary effect of metal dust sprayed with DIME is to kill, injure, or damage by blast without leaving much trace of fragments. When the 1980 Protocol I was adopted unanimously, states did not have such weapons in their inventory; nor did they foresee any conceivable use of such weapons in the future. It could be argued thus that a weapon like DIME if developed under a global NES military project was not prohibited under the protocol, as such micro-shrapnel could still be detectable, no matter how difficult it might be in practice using X-ray methods commonly available to trauma surgeons. Yet, DIME also could be prohibited owing to the seriousness of injury it causes given the minute size of the fragments that are beyond practical surgical manipulation or removal.

A major issue for a specific NES nano military treaty is that nanotechnology, as mentioned, offers not only new propellants and explosives with higher energy density, but also miniature guidance systems for small munitions. Existing arms control treaties such as the United Nations *Biological Weapons Convention* (BWC) could be circumvented by more controllable and precisely targeted nano-weapons delivery agents, such as crewless tanks and pilotless aircraft; partly this would be because such systems would not readily be able to discriminate between combatants or non-combatants or to take responsibility for errors.

Defence strategies based on deterrence would be undermined by stealthy, precision-guided nano-weapons. This would create the need for instantaneous response, creating great potential for accidental war and uncontrolled escalation. Some have argued for international governance to permit molecular shields against nano-weapons, because in their absence nano-weapons could lead to significant geopolitical destabilisation. Thus, although peace and security are major virtues of a global social contract, it is difficult to see nano-weapons contributing to them through deterrence in the same way, arguably, that nuclear weapons have.

7.3 NANOSURVEILLANCE FOR GLOBAL BIOSECURITY?

During one sabbatical period my wife Roza and I lived in London at a house on Pond Street near Hampstead Heath. London, of course, has one of the highest concentrations of CCTV security cameras in any city. At the bottom of the street, opposite the Royal Free Hospital and at the corner with South End Road, was a pizza shop that had once been a bookshop where the writer George Orwell worked. I've always admired Orwell for his works' conscience-driven efforts to warn people against states run by totalitarian regimes. While in London, each time I spotted a security camera I recalled George Orwell's *Nineteen Eighty-Four* and its frighteningly insightful portrayal of a totalitarian state utilising sophisticated technological surveillance methods to control not just the actions, but also the thoughts and emotions of its populace. Some have even said that the Stasi in East Germany, Pol Pot in Cambodia and the dysfunctional regime in North Korea appeared to have used *Nineteen Eighty-Four* as a type of reverse instruction manual on how to dismantle and replace a social contract as a mode of governance. Could this become the dark side of a global NES project focused on military or police aims and outcomes?

On the other hand there can be no doubt that protection of a society against terrorism requires heightened surveillance. My own exposure to such acts involved treating airlifted survivors from the 2002 Bali terrorist bombings while senior registrar in the Intensive Care Unit at the Alfred Hospital in Melbourne. One of the victims included a badly burnt Australian Rules football player, who after treatment and despite skin grafts was able to resume his playing career – as an act of defiance to the terrorists.

The reduction in size of temperature and pressure sensors, listening and spying devices is a continuing global trend in military and police technology. Imagine a global NES project set up to establish a global security system against terrorism – we might call it 'SecuroNet'. This system would be designed to integrate nano, micro and conventional sensors into a national and international network able to detect minute quantities of explosives or biowarfare agents such as anthrax or chemical toxins, to counter terrorist attacks, assist police and security services as well as warn soldiers of battlefield risks.

A global NES civil security project could assist the development of more accurate and covert detection (including imaging, sensors and sensor networks), protection (including decontamination and filtering, personal protective equipment, electromagnetic shielding) and identification (including authentication and counterfeiting, forensics, quantum cryptography). Rather than detecting a pathogen through laboratory testing involving time and resource-intensive PCR, colony count, ELISA and electrophoresis, nanotech biosensors in enzyme, antibodies or nucleic acid in food, water storage and transport, pharmaceuticals and environmental monitoring and agriculture will give rapid, cost-effective warnings.

A global NES security project could offer portable or embedded technology with low power consumption, rapid response times, the ability to recognise many different biological and chemical dangers over a wide range of concentrations, high sensitivity and specificity, speed and integrated systems for remote or embedded monitoring. It could involve ambient sensors at contaminated sites or continuous detection of biological or chemical threats in airports or in clothing of military or security personnel.

A global NES security project might also emphasise the importance of nanoscience applications to national security programs, particularly in relation to improved sensitivity, selectivity and expense of sensors in the military, government buildings, customs areas, hospitals, decontamination locations and schools.

The security policy possibilities created by nanotechnology often appear to be overstated, in terms of both pace and the nature of the changes it may bring to virtually all facets of everyday life. Envision, for example, a global NES security project that involves research by which bacteria or viruses have nanotechnological components included in them that are programmed to take and send images or sounds from within their hosts. Synthetic biology of this type is a nanoscale activity that is no longer in the realm of fantasy. It could involve the design and construction of new biological entities such as enzymes, genetic circuits, and cells or the redesign of existing biological systems using nanoscale bioengineering and chemistry, circuit design and nanocomputing. Would this assist or hinder peace and security under a social contract?

The focus of a global NES project on biosecurity surveillance definitely has points of coherence with a global social contract. The World Health Organization (WHO) International Health Regulations (IHR) are part of an international treaty and have been revised to take into account the resurgence of infectious diseases such as Ebola, Lassa, meningococcal meningitis W135, severe acute respiratory syndrome (SARS) and avian influenza (N5N1). Previously, the IHR covered only the easily quarantined diseases of cholera, plague and yellow fever.

The most significant development in the IHR is the establishment of a comprehensive surveillance system that sets up a public health response framework to deal with international spread of disease, while attempting to retain the balance between security and interference with international trade and travel. The IHR sets up specific criteria for countries in relation to their mechanisms of detecting, reporting and responding to such health threats. An NES global security project could build on governance patterns such as the IHR to incorporate nanotechnology research and development into national and global biosecurity policy.

Yet, the governance aspects of any global NES security project must also consider the potential impact of such heightened surveillance on human rights and ethical principles that are central to the global social contract.

Security nanotechnologies will potentially be responsible for collecting and reporting upon sensitive information about individuals' health status and this raises all sorts of issues for protecting personal freedoms. The right to the integrity of the person, to liberty and security, to the respect for private and family life, to the protection of personal data, to equality and non-discrimination, and to informed consent as outlined in documents such as the ICCPR, the ICESCR, and the UDHR will have to be given due consideration. There will need to be extensive debate as to how best to ensure that basic human rights are not forgotten or violated by a nanotechnology-based security system developed under such a project.

Thus, however important the goals of a global NES security project, they are likely to have significant dual-use problems and risks of misuse. Once again, it is unlikely that international civil society will wish to prioritise expenditure on security uses of nanotechnology when that threatens to put more power into the hands of elites who have caused so much suffering through its abuse and when it would leave untouched competing global health and environmental issues such as poverty, biodiversity loss and environmental degradation.

8. Nanotechnology, climate change and renewable energy

Ye who listen with credulity to the whispers of fancy, and pursue with eagerness the phantoms of hope; who expect that age will perform the promises of youth and that the deficiencies of the present day will be supplied by the morrow; attend ...
— Samuel Johnson, *Rasselas*

8.1 NES GEO-ENGINEERING PROJECT?

I explored the dusty ruins of Nalanda Buddhist University in northern India in the late 1980s. Looking at the toppled and overgrown buildings, it was hard then to imagine that at one time in human history this place briefly had been very special in uniting the best of human wisdom and knowledge with a deep respect for nature. The medieval Chinese pilgrim Hiuen Tsang lovingly described what were then its regularly laid-out towers, forest of pavilions, harmikas and temples seeming to soar above the mists in the sky (so that from their cells the monks might witness the birth of the winds and clouds). He wrote of its azure pool winding around the monasteries and lecture halls, adorned with the full-blown cups of the blue lotus; the dazzling red flowers of the kanaka hanging, while groves of mango trees offered the strictly disciplined and virtuous scholars their dense and protective shade.

The community of Nalanda seems an ideal one – learned human beings attempting to mould their knowledge and conduct according to natural patterns of symmetry and harmony they detected deep in meditation. Yet by another pattern Buddhism was strongly opposed in India by members of the yogic tradition led by Adi Sankara, and by Islamic fanatic Bakhtiyar Khilji, who sacked and destroyed the Nalanda and its library (which burnt to ashes over three months).

In this book we've proposed that humans in our generation have an opportunity to shape the world, this time with the assistance of nanotechnology, according to similar patterns of symmetry and harmony. If the theory expounded here – of legal norms emerging from virtues such as justice, equality and respect for environmental sustainability – is correct, then it is surely appropriate to consider a global NES project focused on remedying problems associated with anthropogenic climate change.

The present rapid increase in the warming of Earth's climate system is on the scientific evidence unequivocally owing to human interference. It is worth reinforcing the relevant basic facts, as they form the backdrop to this call for a global NES project.

According to the United Nations Framework Convention on Climate Change (UNFCCC), the average temperature of the Earth's surface has risen by 0.74°C since the late 1800s and will increase by another 1.8°C to 4°C before the end of this century. The Intergovernmental Panel on Climate Change (IPPC) reports the global sea level rose at an average rate of 1.8 mm per year between 1961 and 2003, but by 3.1 mm per year between 1993 and 2003.

Atmospheric CO_2 has been measured almost continuously at the weather station on Mauna Loa in Hawaii since 1958. Current estimates have atmospheric heat-trapping CO_2 at 385 ppm up from 280 ppm before the Industrial Revolution (and the planetary safe limit set at 450 ppm). Old photosynthesis fuels (coal, oil and natural gas) have a unique ratio of two stable isotopes of carbon (carbons 12 and 13) in atmospheric CO_2, so their abundant presence in the atmosphere confirms industrialisation as one of their major sources.

The International Energy Agency's *Key World Statistics* report that the breakdown of energy sources across the world is heavily dominated by old photosynthesis fuels: oil 34.4 per cent, coal 26 per cent, natural gas 20.5 per cent, combustible (wood) and renewable (solar, wind) 10.1 per cent, nuclear 6.2 per cent and hydro 2.2 per cent. Half the world's atmospheric CO_2 contributed by the industrial uses of the first three of these energy sources is removed by oceans. In sea water CO_2 forms carbonic acid (H_2CO_3) making it more acid and thus disrupting oceanic ecosystems and food chains. Sea animals, for example, find it harder to create the shells they need from carbonate. Amongst the oceanic organisms likely to be damaged are phytoplankton such as coccolithophorids that float near the ocean surface and the symbiotic algae that protectively coat coral.

The IPCC notes that most of the verified increase in globally averaged temperature since the mid-twentieth century is likely owing to the observed increase in anthropogenic (that is, human) greenhouse gas concentrations (particularly carbon dioxide, methane and nitrous oxide). This research shows that, even if greenhouse gas concentrations were immediately stabilised, anthropogenic warming and sea level rise (and their associated social and ecosystem damage) will continue for centuries owing to the long timescales associated with climate feedback systems. The likely amount of temperature and sea level rise, however, will vary greatly depending on the intensity by which we continue to burn old photosynthesis fuels (oil from cyanobacteria in shallow oceans and coal from decomposition of much more widespread forests) as our main source of domestic and industrial energy.

If nothing much changes, world temperatures could rise by between 1.1°C and 6.4°C (2.0°F and 11.5°F) during the twenty-first century and sea levels by 20 to 60 cm. The probability of exceeding 2°C rises to 53–87 per cent if global greenhouse gas emissions are still more than 25 per cent above 2000 levels in 2020. As the atmosphere heats, it will contain more water vapour resulting in more frequent droughts, floods and extreme weather events (including cyclones). The destruction of food chains in oceans as a result of acidification may become an even greater problem for humanity than resultant severe droughts, floods and other adverse weather events.

The United Nations predicts the world population will grow from 6.5 billion in 2006 to over 9 billion by 2050. Demand for old photosynthesis fuels will correspondingly increase in developing countries including China, India and the Middle East. The International Energy Agency predicts that world energy demand will increase by 45 per cent between 2006 and 2030.

A global NES project in this area could take its lead from the UNFCCC and have as a stated primary objective the 'stablisation of greenhouse gas concentrations in the atmosphere at a level that would prevent dangerous anthropogenic interference with the climate system'. Such a global NES project might likewise require remedial policies including those using new technologies to 'ensure global benefits at the lowest possible cost' and cover all relevant sources, sinks and reservoirs.

This global NES climate change project could, like the *Kyoto Protocol*, have at its governance core an international treaty to reduce dependence on old photosynthesis fuels with the aim of slowing global warming by decreasing the carbon intensity of global energy supplies. Its protocol might include a mechanism whereby nanotechnology would assist industrialised nations to systematically reduce greenhouse gas emissions (for example by decommissioning old coal-fired electricity power stations and introducing green technology in a gradual shift to 'zero-carbon' energy sources).

Unfortunately, the supranational corporations controlling the old photosynthesis fuel industry (and those in governance positions beholden to it) have sponsored and promoted policies allowing for only a gradual response to climate change, with the largest reductions in emissions delayed into the future. These include the costs of scrapping embedded carbon-intensive capital, the time required to roll out cost-effective, decarbonised green technologies and the lower net present value of future mitigation costs when positive discount rates are used.

So what type of specific solution could a global NES project focus on to resolve this problem? Let's leave aside alternative suggestions that humanity should fund a global scientific effort on space travel (for example using plasma-based space engines or nanotechnology-based solar sails); maintaining the Earth as an oasis of life that facilitates our rapid colonisation of other worlds in the 'goldilocks zone' near our own sun ('not too hot and not too cold' and with stable liquid water exposed to photons). This certainly represents an important challenge for our ingenuity and creativity, but like many of the alternatives previously canvassed it does not significantly and immediately address what must surely be a more pressing ethical priority under a social contract and science-based conceptions of natural law – to reduce loss of life and suffering in our immediate surroundings.

Such arguments would not apply so directly to an NES 'geo-engineering' project attempting to create global solutions to anthropogenic climate change. One illustrative example of geo-engineering projects involves installing a fleet of mirrors in geostationary orbit to strategically reflect the sun's rays. Use of nanoparticles would vastly increase the reflective surface area of such orbiting shields or mirrors. The cost of launching the materials into space would be great, but it purportedly provides a predictable impact on atmospheric albedo (reflectivity).

A related NES geo-engineering option might focus on dispersing, via commercial airliner fuel, nanosulphur dioxide aerosols, dust or highly reflective nanoparticles into the stratosphere (for longer duration of effect and less impact on ecosystems) to increase planetary albedo. The aim once again would be to promote planetary cooling, though the nanosulphur might turn the sky yellow-white instead of blue, with deep red sunsets. Indeed, Edvard Munch painted the background to *The Scream* after witnessing the sunsets in Oslo caused by atmospheric particles released after the eruption of Mt Krakatoa in Indonesia. Volcanoes in Iceland (in the Northern Hemisphere) and South America (in the Southern Hemisphere) have recently had a similar effect.

Proponents of such an NES geo-engineering project claim that volcanic eruptions (like Mt Pinatubo on the Philippine island of Luzon in 1991, which injected 20 megatons of sulphur dioxide into the atmosphere) provide case studies that global cooling does follow global dust clouds. Indeed the geologic record seems to link such eruptions (through their diminution of sunlight and global photosynthesis) with catastrophic disruption of ecosystems. They assert that an albedo change of about 1.5 per cent is required to offset the impact of a doubling of atmospheric CO_2.

An NES geo-engineering project on the Earth's surface might, for example, use iron nanoparticles to seed massive algal growth in oceans to absorb excess atmospheric CO_2. A governance aspect of such a project might involve a supranational corporation setting prices for tax-exempt donations to facilitate ocean fertilisation using iron nanoparticles to promote swathes of carbon-soaking phytoplankton – these would in theory absorb excess atmospheric carbon dioxide then sequester it in the sea depths as they die.

Such a global NES geo-engineering project could build on at least ten scientific and government-supported efforts to 'seed' sections of the world's oceans with iron nanoparticles. Some examples included the efforts by the US, the UK and Mexico off the Galapagos Islands (IRONEX I and II), the Southern Ocean Iron Release Experiment Expedition (SOIREE) in 1999 involving New Zealand, Australia, Canada, the UK, Germany, the Netherlands and the US, Japan's Subarctic Pacific Iron Experiment for Ecosystem Dynamics Study (SEEDS) and the 2002 Subarctic Ecosystem Response to Iron Enrichment Study (SERIES).

In 2007, two private companies, Ocean Nourishment Corporation of Australia and Planktos, Inc. of the USA, were prevented from carrying out ocean fertilisation activities in the Sulu Sea and near the Galapagos Islands. They planned to scatter nanoparticulate iron sulphate particles over 300 square kilometres of open ocean in the Scotia Sea close to Antarctica to provoke a massive plankton bloom that would facilitate carbon sequestration.

Critics might claim that such a global NES project is really all about helping supranational coal and oil companies buy carbon credits, that the carbon removed to the ocean might dangerously acidify it or that other major impacts on ocean food chains would be initiated. A host of civil society NGOs are already calling on governments backing the plan to implement and respect an international moratorium on ocean fertilisation.

A related geo-engineering solution involves dumping billions of tons of quicklime into the oceans so the excess atmospheric CO_2 can be sequestered in limestone. Unfortunately 2.7 gigajoules of energy will be required to convert each ton of limestone. That's the equivalent of 10 billion barrels of oil. Such a project also risks significant damage to ocean food chains.

Yet for developed nations (whose industrialisation created the climate change problem) to experiment this way on the world's climate without United Nations approval when the risks are unknown, potentially catastrophic and potentially irreversible could seem to international civil society a highly divisive and provocative act. It may even be contrary to the United Nations *Environmental Modification Convention* that prohibits the use of climate as a weapon against other states. This convention targeted acts such as the CIAs 'Project Popeye' in the Vietnam War whose object in cloud seeding between 1966 and 1973 was to make the Ho Chi Minh trail impassable and to drown out North Vietnam's rice crop (even to disrupt the anti-war protests of Buddhist monks).

In the early 2000s Craig Venter (who led the private sector component of the macroscience project to map the human genome) sailed his 90-foot yacht *Sorcer II* into the Sargasso Sea in search of new marine microbes that he could genetically engineer to soak up CO_2 and become super-photosynthesisers across the oceans of the world. Others propose placing an oily film over the oceans (some oil companies may already have indirectly supported such an approach) to restrict evaporation and reduce hurricane development. Still others suggest solar power stations in stationary orbit around the Earth that are capable of heating the air around a hurricane and mollifying it.

To show how complicated the debate over global governance in this area has become, James Lovelock, the originator of the Gaia hypothesis, has proposed that the most practical solution to the warming of the Earth from excess carbon dioxide production is to build nuclear power stations across the globe. This solution, apart from being excessively energy and water intensive to construct and run, centralises power generation in corporate hands, increases the risk of weapons proliferation and radiation damage to the environment and human populations.

Our previous discussions have emphasised how important the precautionary principle is to ensuring safe and cost-effective use of new technology in a global social contract supported by science-based natural law. Imagine being on a ship in the sea where nanoiron had been sprinkled as part of a geo-engineering NES project. The clear blue waters are now green owing to a massive growth of phytoplankton. But how sure are we that these organisms will not only absorb atmospheric CO_2 but also sequester it in the ocean depths? To actually absorb a huge amount of CO_2 from the atmosphere a large area of ocean would need to be fertilised. Would that, as algal blooms do in rivers, then deprive the surface water in that area of oxygen, killing marine life crucial to its food chain? Proponents of this type of global NES project claim they need 360 ships running full-time to fertilise 46 million square kilometres of ocean, costing between $10 and $100 billion per year.

Wouldn't that money be better spent on nanotechnology-based solar technology? In any event, would it stop other critical problems of oceanic ecosystem integrity, such as those associated with nitrogen-based fertilisers and hormone-changing chemicals leaching into the oceans, over-fishing by commercial operators, and disruption of the ocean food chain by killing large species such as sharks?

Parties to the UN *Convention on Biological Diversity* (CBD) and the *London Convention of the International Maritime Organization* – the treaty that governs the dumping of wastes at sea – have already agreed that no ocean fertilisation activities would proceed until there was 'an adequate scientific basis on which to justify such activities, including assessing associated risks', and until 'a global, transparent and effective control and regulatory mechanism is in place for these activities'. So far, there is no such mechanism.

Governments in developed countries may push for a geo-engineering NES project because it represents a 'technological fix' to climate change that allows lucrative industrial opportunities based on atmospherically polluting fuels from old photosynthesis (such as oil and coal) to continue undisturbed. In economic terms, geo-engineering offers to convert the problem from one requiring immediate, globally coordinated regulation to mitigate climate forcing and behaviour modification in developed nation consumers away from profligate energy use towards more politically simplistic technology innovation, transfer and international cost-sharing.

Yet global governance mechanisms already seem to be reaching a consensus that an NES geo-engineering project would not be in accord with the basic virtues, principles and laws of a global social contract. Nearly 200 scientists from 14 countries met a few years ago at the Asilomar retreat centre outside Monterey, California. This replicated a prior meeting at that site in 1975 in relation to the virtues and principles for the safe study of deadly pathogens. Their five-day meeting attempted to scope how geo-engineering (much of it nanotechnology-based – such as cloud-brightening and ocean algal blooms) might sit within the ethics, law and human rights of a global social contract. They found it did not.

A major problem with garnering public and policy support for an NES geo-engineering project is that it focuses on remedying problems associated with human overpopulation and energy demands without addressing their root causes. An NES geo-engineering project appears to conform to a business-as-usual model for the old photosynthesis-fuelled industries running corporate globalisation. It is unlikely to advance a broader global social contract agenda by underpinning principles such as that power production should be decentralised, available equitably and should facilitate environmental sustainability.

8.2 NES PROJECT REDUCING OR SEQUESTERING CARBON EMISSIONS AT SOURCE?

When I studied medicine at Newcastle University by the coast north of Sydney in New South Wales I lived for a period in a group house in the industrial suburb of Mayfield. At night the illuminated smoke stacks from the steel works could be seen releasing their emissions upon the sea winds over the lines of well-lit ships waiting to enter the harbour and load with coal. The story used to go that you could tell when it was knock-off time at the foundries because the air quality inspectors went off duty and plumes of foul-smelling gas rose into the air.

Every year vast amounts of volatile organic compounds (VOCs) and carbon dioxide pollute the atmosphere from the waste gas streams of such industrial plants. Globally, approximately 25,000 million tonnes of carbon dioxide a year are so released (6,500 million tonnes of carbon). A negative policy impasse called 'carbon lock-in' has occurred as the governments of developed and developing economies are forced (chiefly because of their economic significance) to perpetuate old photosynthesis-fuelled infrastructures such as petrol-based automobile transportation networks and coal-fired electricity systems, despite known negative environmental impacts.

The Copenhagen Accord specifically recognises the consensus scientific view that the increase in global temperature should be kept below 2°C and supports a theoretical cap on emissions, although the Accord does not dictate how the cap translates into national emissions allocations. The Accord does not pre-empt the means by which emissions reduction commitments are to be achieved. This leaves open the use of governance mechanisms such as a cap-and-trade scheme, a carbon tax or other types of 'direct action'.

An NES project could focus on sequestering greenhouse gas emissions at source. This would cohere with the fact that the governments of the rapidly growing economies in China, India, Japan and Indonesia now have policies to significantly reduce emissions intensity of industrial output or overall greenhouse gas emissions by 2020. Mechanisms for achieving these goals include decommissioning inefficient coal-fired power and manufacturing plants, investment incentives for low-carbon energy, taxes on old photosynthesis fuel extraction, trials of carbon market mechanisms, reduction of GDP growth targets and supporting the United Nations REDD Programme (Reducing Emissions from Deforestation and Forest Degradation).

An NES greenhouse gas sequestration project, in governance terms, could be considered an extension into the new technology policy of the

1972 Stockholm Conference on the Human Environment that helped establish the United Nations Environment Programme (UNEP) with Earthwatch and the Global Environment Monitoring System (GEMS), as well as the International Referral System for Sources of Environmental Information (INFOTERRA) and the international Registry of Potentially Toxic Chemicals (IRPTC).

The easiest way of reducing such emissions might be to reduce demand for old photosynthesis-fuelled electricity. Global promotion of nanotechnology-improved home insulation could be a highly effective and very simple energy-reduction strategy. It reduces the need for both heating and air-conditioning power. In Australia, for example, policy-makers rolled out home insulation installation with subsidies on a national scale. Yet insulation (however valuable and made more efficient by nanotechnology) can't be viewed as more than a supplement to a more complete renewable energy solution.

There are many other examples of the types of nanotechnology that could be developed for global use under such a global NES greenhouse gas emissions reduction project. Skyonic, for example, is a company developing nanotechnology to capture carbon dioxide emissions from power plants, mix it with sodium hydroxide and produce high grade baking soda. Nano-fibres with higher membrane area-to-volume ratios (>3000 m^2/m^3) than existing extractors have been developed by researchers to extract these VOCs and reuse them for food, beverage and dairy processing, chemical, pharmaceutical and oil/gas processing. These nanofibres are more cost-effective than conventional thermal oxidation pollutant-trapping processes requiring less than 5 per cent of their energy.

A problem that such an NES project could address is that major atmos-pheric pollutants, as well as greenhouse gases also include toxins such as methylmercury (MeHg). Methylmercury is poisoning our nervous systems globally predominantly from coal-fired power station emissions via ocean deposition and accumulation in seafood. Most current monitoring of levels of such pollutants utilises large, fixed devices that provide localised data, or require an operator to go to a location, take a sample and have it analysed in a laboratory.

A global NES project in this area could produce nanosensors distributed widely, and automatically relaying this information back to a central control. An example developed at Ispra, Italy, is based on nanostructured thin films of tin oxide. It is a solid state sensor able to detect nitrous oxides and carbon monoxide with a faster response time, being cheap and re-usable. Similarly, a global NES project might facilitate a company develop-ing cerium oxide nanoparticles to increase petrol efficiency.

Another illustrative potential output from such a global NES project could involve old automobile tyres. Each year over 600 million tyres reach the end of their life cycle in Europe and North America. Many global governance initiatives have now banned scrap tyres, shredded tyres and tyre residues from being landfilled or stockpiled. Yet a specific nanotechnology-based technique could reduce 40,000 tonnes of carbon dioxide greenhouse gas emissions per used tyre facility. This is how. Pyrolysis of scrap tyres has been commercially unsuccessful mostly because of the poor quality of the carbon rich pyrolysed by-product. Yet nanotechnology has allowed a practical method to post-treat the pyro-carbon into market products: the crude pyro-carbon char is refined, upgraded and blended with standard commercial carbon black grades into a commercial substitute for carbon black used in the rubber and plastics industries. Nanotechnology, additionally, could assist in reducing levels of environmentally damaging black carbon (a component of soot from incomplete combustion of coal and wood and a significant greenhouse gas).

The main problem, however, with any global NES project primarily focused on carbon sequestration would be the extent to which it aims at remedying problems associated with human overpopulation and energy demands without addressing their root causes.

8.3   NES PROJECT INCENTIVISING RENEWABLE ELECTRICITY?

One of the most pleasurable aspects of living in Canberra, apart from the easy drive east to Depot Beach on the south coast in summer, is the similarly short trip south to the Snowy Mountains for cross-country skiing in winter. One hour out of Canberra, heading east, is the township of Bungendore with its excellent second-hand bookshop and nearby huge wind farm. Travelling south and passing through Michelago and Bredbow on the Monaro plains, one comes upon Cooma, the home of the Snowy Mountains Authority hydroelectricity scheme. Just after the Second World War large numbers of European migrants assisted in building tunnels through mountains (with much loss of life) that created this vast hydroelectricity project. It still operates and was only recently saved from privatisation. Both these renewable energy schemes (the Bungendore wind farm and the Snowy Mountains Authority hydroelectric scheme) are now generally regarded as valuable parts of the natural environment as well as important parts of the puzzle of our social future once we have made our way through the crisis posed by our transition away from dependence on old photosynthesis-based fuels.

In 2001 the population of Earth was approximately 6 billion and world gross domestic product (GDP) about $7,500 per capita. By 2050 global population is projected to increase to over 9 billion and GDP to about $15,000 per capita. The use of old photosynthesis fuels to satisfy the consequent growth in energy demand will become increasingly problematic owing not only to limitations in supply and geopolitical location, but also to their proven contribution to anthropogenic climate change. The policy push will be to de-carbonisation of energy. This can be characterised as a transition from fuels with low to high hydrogen/carbon ratios on the road to a hydrogen economy dominated by renewable energy sources. A global NES project could prompt world policy-makers towards such developments.

A global NES renewable energy project would directly support the United Nations *Millennium Development Goals* (MDGs). The MDGs can be regarded from the science-based natural law perspective advocated here as an authoritative attempt at expression of a global social contract (from a 'positivist' conception, of course, they do not have the status of international law according to the 'rules of recognition' in Article 38 of the *Statute of the International Court of Justice*). The MDGs involve agreed targets set by the world's nations to reduce poverty by 2015. The most relevant to renewable energy are goal number 7: Ensure environmental sustainability, goal number 8: Develop a global partnership for development and Target 18: In cooperation with the private sector; make available the benefits of new technologies. The United Nations MDGs can be viewed as setting ideals for government policy to gravitate towards – increased education and health care, as well as reduction of poverty and childhood and maternal mortality rates. Most relevant models predict that without equity in global access to electricity it is unlikely that such MDGs will be achieved.

At present about 3 billion people on this planet rely upon intermittent and unhealthy supplies of wood, cow dung, coal or kerosene for domestic cooking and heating. Cooking on open fires and in insufficiently ventilated homes with wood and coal-based fuels causes approximately 2 million premature deaths every year, as well as many more burns and respiratory diseases (disproportionately amongst women and children). Women in particular also suffer from long travel to collect such fuel for heating and cooking, plus lack of domestic lighting for education. This further compounds the world's population problems, as there is now good evidence of an inverse relationship between birth rates and parental education. The global governance issue here that a global NES renewable energy project would have to address is equitable access to electricity for approximately 800 million households by 2030.

The International Energy Agency (IEA) has concluded that deep cuts in anthropogenic greenhouse gas emissions are possible, but to achieve large-scale reductions by 2050 renewable energy sources such as solar and wind will need to become as large a component of total energy outputs globally as the coal sector is today. Atmospheric carbon dioxide levels already have been significantly reduced in some nations through the adoption of non-old photosynthesis means of generating electricity. They were reduced by 40 per cent in France between 1996 and 2007, for example, owing to adoption of nuclear power and lowered by 30 per cent over the same period in Denmark especially through the enhanced adoption of wind power.

A macroscience NES project focused on renewable energy could emphasise the role of nanotechnology in developing products that will reduce household use of solid (biomass) fuels; generating cleaner electricity from renewable sources (such as solar, wind, ocean, geothermal and hydroelectric power) and improving household energy efficiency (for instance by providing efficient heating and cooling appliances, improving home insulation). Such aims fulfil the normative criteria outlined earlier as drawing upon a global social contract and science-based natural law. Providing universal electricity access, however, poses critical challenges to equitably produce, deliver, install, manage, operate and maintain the NES solutions involved.

In 2010, for instance, at the same time as pressure was mounting in many developed nations for a tax upon atmospheric emission of carbon dioxide, a study by the IEA revealed that the world economy spends more than $550 billion in subsidies that encourage consumption of 'old photosynthesis' fuels (oil, gas and coal). Such subsidies reduce global energy security, impede investment in clean energy systems, undermine efforts to deal with the threat of climate change and restrict the capacity of governments to provide education and health care services. Imagine if a global NES project were able to create products that assisted government policy to systematically replace these outdated and environmentally deleterious subsidies with those supporting renewable or sustainable energy solutions.

The governance aspects of such an NES project might facilitate policies to set renewable energy standards for home lighting, heating and appliances, reduce transaction costs, align incentives, monitor performance and overcome market failures. They could aim to develop international standards to promote private sector investment that drives down the roll-out costs of efficient renewable energy nanotechnologies. They might promote universal domestic access to basic green energy supplies that would improve the quality of air in the cities of many developing nations. They could work towards integrating steady state economics models into industry policy, replacing the illogical policy hallucination that endorses the unsustainable prospect of endless economic growth.

Let's examine some illustrative examples of potential global NES renewable energy for electricity projects. Hydroelectric dams would be a potential candidate (for example by enhancing the efficiency of turbines or power transmission). Hydroelectricity has the additional advantages of regulating river flows to minimise flood damage, providing a source of farm irrigation and urban drinking water, as well as being a potential source of water for steam generation (for instance in a nuclear power plant). Large hydroelectric dams are situated on about half the major rivers of the world and provide about 10 per cent of the world's total electricity needs – often to large industrial users such as aluminium smelters and paper mills. The constantly turning turbines can provide backup for rapid supplies to meet unexpectedly large demands in the electricity grid.

A variation of this approach is pumped hydroelectricity, which uses new generation photovoltaic (PV) technology to pump water up approximately 400 m to a holding dam where it can be run down through turbines (thus solving the problem of intermittent energy from PV).

Yet, the hidden cost of hydroelectric dams (were they to be the focus of a global NES project) is significant and relates to factors such as the number of persons and animals displaced, farms and communities damaged and ecosystems destroyed. The biomass in the flooded areas gradually decays, releasing greenhouse gases. The replenishment of soil and of water storages downstream is halted or reduced. With limited options now available for such large-scale projects, the World Bank has almost ceased its lending for them. Such factors make it unlikely that hydroelectricity will become the centrepiece of a global NES project.

Considerable debate surrounds whether to term nuclear fission or nuclear fusion sources of renewable energy for electricity generation. Claims (such as those by James Lovelock) that nuclear fission electricity generation provides the carbon-neutral solution to our global energy and greenhouse gas problems are tempered by the considerable radiation safety risks associated with disasters such as those at Three Mile Island, Chernobyl and Fukushima, as well as the potential contribution of the necessary uranium enrichment program to military weapons proliferation and terrorist use. Such factors have been part of the reason that no new nuclear fission electricity power plants have been commissioned in the US since 1978 and that policies are in place for their phasing out in Belgium, Germany, Italy, the Netherlands, Spain and Sweden.

Hidden costs of nuclear fission power include the intractable problem of safe long-term storage of nuclear waste (even in allegedly geologically stable areas), the amount of materials and carbon-intensive energy required to build such facilities, the long time needed to obtain necessary approvals, their lack of widespread public support (especially when located near cities) through fears of radiation contamination and the amount of water they use. There is also the issue of their vulnerability to and severe national consequences of terrorist or military attack. Nuclear fission for electric power is unlikely to be the main focus of a global NES project striving for governance coherence with a global social contract.

The International Thermonuclear Experimental Reactor (ITER) involves a $10 billion collaboration of researchers from the US, Europe, Russia, China, Japan and Korea to construct a fusion reactor. Could a global NES project centre on nanotechnology facilitating nuclear fusion and should it resemble the ITER endeavour? A major problem for ITER (particularly in terms of coherence with a global social contract) is that like hydro, nuclear, ocean and wind power, to be cost-effective it has to involve large corporate-owned, utility-scale constructions that integrate into the existing electricity grid. Electricity generation from nuclear fusion, in other words, is a large-scale solution requiring extensive distribution networks and huge initial capital costs and technology maintenance. It will not immediately assist the energy problems of those dying of poverty in developing nations and so does not manifestly cohere with science-based natural law, or most rational conceptions of a global social contract.

The wind power industry is increasing by 25 per cent per year and annual wind turbine sales total approximately $10 billion globally. Yet to be cost-effective wind farms must be large and they often meet significant local community resistance based on allegedly negative factors such as sound and landscape appearance (including impact on the tourist industry). Offshore wind farms meet fewer objections, but still raise concerns about impacting on fish migrations and causing boats to make detours. Further, the intermittent nature of wind means that the resultant electricity levels fluctuate considerably (the turbines must shut down in high winds). For such reasons it is unlikely that wind power will be the focus of a global NES project.

Geothermal, ocean and fusion energy also do not appear to be stand-out candidates for a global NES project focused on renewable energy. Geothermal is a significant renewable energy option for nations (such as Iceland, Bolivia and Philippines) that can tap the temperature differential of water in geothermal reservoirs. Unfortunately, such locations are limited, poisonous hydrogen sulphide gas is produced and the technology is unproven for use globally. The movement of oceans as tides and waves does promise a reliable, constant and cost-effective form of base load electricity, but this is limited by the need to find appropriate locations and the requirement to maintain sophisticated equipment.

NES solar electricity processes, such as Dye Solar Cell technology, have the potential to produce electricity which is 'grid competitive', that is at the cost to the consumer of electricity from old photosynthesis fuel power stations. Currently 90 per cent of the solar cell market is based on crystalline silicon wafers, with thicknesses of 200–300 μm. Dye Solar Cells (DSC) nanotechnology was developed by Professor Michaël Grätzel in Switzerland in 1991. The excellent price/performance ratio of these cells makes them likely to have an important role in low-cost, large-scale solutions for renewable energy. Nanotechnological improvements in PV include those developed from research such as that by Kylie Catchpole at the Australian National University in which nanotechnology-based plasmonic solar cells greatly increase the efficiency of photon harvesting.

There has been a rapid growth in worldwide solar electricity (PV and solar thermal) and it is predicted that soon parity will be achieved worldwide between the cost of power from such sources and from the general electricity grid. One model is that with affordable solar electricity readily available vehicles will begin to run off electricity (for instance using improved batteries).

Yet, although PV efficiencies can be improved with nanotechnology, even if the majority of houses in the future, for instance, have 10 m^2 of 20 per cent efficient solar panels per occupant, then only ~5 kWh per day per person will be produced. Even large solar farms (for example taking up 200 m^2 per person with 10 per cent efficient solar panels) could produce an average of ~50 kWh/day. This is still a long way short of ~125 kWh/day average consumption and still leaves the problem of power storage for night use and transport.

The hydrogen economy has often been mooted as the central direction of renewable energy efforts and this looms as a potential flagship component of a global NES renewable energy project. Indeed, it is possible to plot humanity's necessary global fuel transition as one moving from sources such as wood (ten atoms of carbon per atom of hydrogen), coal (two carbon atoms per hydrogen atom), oil (one carbon atom per two hydrogen atoms) and natural gas (one carbon atom per four hydrogen atoms) towards energy from pure hydrogen.

A global NES project, in other words, could focus on overcoming major technical stumbling blocks to the hydrogen economy. Hydrogen is the commonest element in the universe, but in its usable form (H_2) (for example to make ammonia or refine petroleum) it currently must be created by chemical reformation of natural gas, coal and oil. Fuel cells are being developed that combine hydrogen with oxygen to generate electricity producing heat and water as valuable by-products. Yet unresolved major problems include the requirement to store hydrogen under pressure to fit in vehicles and buildings, the lack of a cheap, spatially and temporally abundant as well as non-energy-intensive source of hydrogen and the development of fuel cells robust enough to cope with the stresses of driving vehicles in all road conditions.

The funding and administration of an NES global renewable electricity project might draw on existing programs such as the Renewable Energy and Energy Efficiency Fund (REEF) as well as the Climate Investment Funds of the World Bank and other Development Banks, the Global Environment Facility (GEF), or the Climate Investment Funds (CIF). Such a project could promote investment and international standards on technology transfer and energy saving. The increased energy efficiency of NES devices could free up capital resources for investment. If that happens then such a global NES project might be advertised as allowing for instance the IEA's recommended threshold of 100 kWh per person per year to be reached without necessarily reaching the predicted 1.3 per cent increase in greenhouse gas emissions above current levels.

An NES universal electricity access project could be promoted to the public as an example of nanotechnology assisting governments to reach the United Nations' MDGs, address anthropogenic climate change, promote steady-state economic prosperity and conserve critically threatened natural resources and biodiversity. A global NES renewable electricity project would thus receive a high policy profile as assisting National Appropriate Mitigation Actions (NAMAs), Low Carbon Growth Plans (LCGPs) and policies for phased introduction of low-GHG emitting technologies.

At the level of international governance, an NES renewable electricity program might facilitate equity of access and improved energy efficiency being integrated into all relevant programs and projects of the United Nations system and those of member states through technical and financial support, knowledge networks, targeted, technical interventions, and donor projects. The UNEP-led Global Network on Energy for Sustainable Development (GNESD) provides a good example of such an approach. The governance mechanisms of a global NES renewable energy project could draw upon national or international Renewable Energy Future Funds to assist governments to reach carbon-reducing and renewable energy targets.

Associated NES project policy mechanisms at the domestic level might include investment subsidies of fixed government payments (cash up front, tax deductions or in the form of tradable certificates) per kWh for renewable energy input into the grid (such policies for example triggered very fast development of wind energy in Germany). Other mechanisms might include permitting electricity wholesalers to offer citizens the capacity to pay a premium to source their household energy from renewables, or to get a reduction on their bill proportional to their input to the grid (for instance from solar panels).

Orthodox renewable energy law and policy emphasises the alleged superiority of so-called 'free market' mechanisms over direct government subsidy, for instance in the form of feed-in tariffs. This preference is ideological and misguided, and is largely corporate lobby driven. It rests on the assumption that 'market-based' mechanisms inevitably, efficiently and rapidly will deliver environmental solutions at 'least cost'. Moreover, it fails to acknowledge that the problem of uncontrolled greenhouse emissions was caused by corporate-initiated market failure in the first place. As mentioned, large amounts of direct energy and transport subsidies have for decades propped up the revenue streams of corporations dominating old photosynthesis fuel production and consumption, far in excess of those supporting renewable energy and energy efficiency. The effect of these subsidies is to make the cost of old photosynthesis fuel energy artificially low and to indirectly make renewable energy alternatives less competitive.

Fostering technological innovation within the power generation sector is often pointed to as a key to 'decarbonising' the global economy. Hence, the governance components of a global NES renewable energy project might include a treaty creating international law obligations on governments to pass renewable energy support laws to compensate NES renewable electricity generators for avoided external costs (for instance carbon dioxide pollution) and provide external benefits in terms of environmental protection, technological innovation, and employment, as well as preventing 'lock-in' to existing technologies that creates barriers to market entry for potentially low-cost alternatives. Such statutes and policies would be required by the same global NES treaty to be combined with measures to prevent fraud and corruption. As mentioned earlier, such governance approaches could include variations of the US false claims *Qui Tam* legislation that accords whistleblowers on corporate fraud 15 to 30 per cent of the triple damages awarded to a government after a successful case.

Yet, a diversity of renewable energy inputs into the electricity grid (for instance wind, biomass, tide, geothermal, hydroelectric, solar PV and solar thermal) already is being developed. Arguably this process would not benefit much from the enhanced public and policy profile of being accorded the status of a macroscience NES project.

In conclusion, most of the above renewable energy candidates for a global NES project are predicated on the continued existence of an electricity grid with the attendant corporate control and privatisation problems (exemplified in the Enron scandal where an electricity grid was manipulated like an unregulated stock exchange) as well as the significant costs of maintenance. This does not cohere as well with science-based natural law and the global social contract models that are arguably more coherent with 'off-grid' more presumptively democratised household and community-based energy solutions. There is an area of nanotechnology research capable of providing such a viable alternative and it will be explored in the next chapter.

9. Nanotechnology's moral culmination: a Global Artificial Photosynthesis project

> A photon of light travels 150 million kilometres
> to reach the earth's surface in about eight minutes
> Yet it takes a plant seconds to capture its energy, process it and store it in a
> chemical bond ...
> Photosynthesis: the plant miracle that daily gives us bread and wine,
> the oxygen we breathe, and simply sustains all life as we know it.
> – David Beerling, *The Emerald Planet*

> There will be just bare awareness paired with its preoccupation in the present.
> This is something with no sense of 'inside' or 'outside' –
> a condition whose features are peculiar to the mind itself.
> It is as if everything has undergone a revolution.
> – Ven. Ajahn Thate, *The Autobiography of a Forest Monk*

9.1 GLOBAL ARTIFICIAL PHOTOSYNTHESIS: OUR GREAT SCIENTIFIC CHALLENGE

In 2010 I visited Namibia for a workshop about the development of African components to the UNESCO global bioethics and health law database. After the meeting, I drove my wife and son around the country in a four-wheel drive. Amongst the interesting areas we visited was Zebra River Canyon. Here you are able to view ancient fossil stromatolites (some of the earliest photosynthetic organisms) and see, in open fields, Stone Age hand tools in the same positions in which they were made or discarded.

During our stay at Zebra River I was reading Frank Wilczek's *The Lightness of Being* (about how the subatomic world can be viewed as light governed by the laws of quantum physics). At night we watched shooting stars rain over a landscape that had changed very little in hundreds of thousands of years. Such experiences seem to compress the evolution of human civilisation and its technological progress into a unifying historical context with the fragile success of life in colonising this planet.

At this time I had been awarded an Australian Research Council Future Fellowship to study how nanotechnology might be governed to help resolve some of the great public health and environmental challenges facing humanity. A Brocher residential scholarship at Hermance by Lake Leman outside Geneva had allowed me to consider this problem at depth. With the assistance of Emeritus Professor John White from the Research School of Chemistry at the Australian National University I had conducted light scattering and neutron scattering experiments on nanoparticles at the Australian ANSTO facility at Lucas Heights and also at the Laue-Langevin Institute at Grenoble. As we drove to Grenoble I remember saying to Professor White: 'Energy, food, water and climate change are the big global issues of our time; the solution has to have something to do with hydrogen and carbon dioxide. That's what nanotechnology research needs to focus on.' The identity of that solution wasn't yet clear to me, but I increasingly suspected it had something to do with photosynthesis.

The turning point came when I decided to include 'artificial photosynthesis' in the title for an oral presentation at the Nanotechnology for Sustainable Energy Conference sponsored by the European Science Foundation on 7 July 2010 at Obergurgl, Austria. That conference was run on what is called the Gordon style with delegates having afternoons off to, in this case, walk through the forests and boulders along the foot of the glacier. On one such afternoon I met Professor Peidong Yang from the University of California, Berkeley. Peidong told me he was awaiting the result of a multi-million US Department of Energy grant to study artificial photosynthesis. 'When do you think you can produce a marketable device?' I asked. 'In my lifetime,' he said. 'Well, you're a youngish sort of bloke,' I replied, 'I don't think the planet has got that long. What about a Global Artificial Photosynthesis project, like the Human Genome Project?' 'Good idea,' Peidong responded, then added, 'Why don't you do something about it?'

That evening I reconfigured my oral presentation to make it a call for a Global Artificial Photosynthesis (GAP) project. That was the central theme of the talk I delivered the next day. I found that the 15th International Congress of Photosynthesis was meeting the following month in Beijing. The date for abstracts had closed, but I wrote an abstract anyway entitled 'A Global Artificial Photosynthesis Project: Seven Models and Seven Reasons Why', then emailed it to every senior participant at the conference. Of course, this was considered somewhat brash, even characteristically Australian, some replied. Nevertheless, Professor Graham Farquhar from my university, who was chairing a session on global photosynthesis, managed to give me a seven minute oral speaking spot. Back in Australia I discovered that a colleague Warwick Hillier was editing a volume for the Royal Society of Chemistry on solar fuels. It was almost ready to go to press, but a few contributors had pulled out or were slow. I offered to write a chapter scoping a Global Artificial Photosynthesis (GAP) project and wrote it for him in two weeks. It was accepted.

This book opened by emphasising what should be blindingly obvious to global policy-makers: the importance of photosynthesis to the sustainability of ecosystems and human civilisation. It's time for the coda. Photosynthesis, the ultimate source of our oxygen, food and old photosynthesis fuels (including oil (from decayed cyanobacteria in shallow oceans), and coal and natural gas (from decomposed old forests)), has been operating on Earth for 2.5 GYr, since a time known in geological circles as the great oxidation event (GOE). Photosynthetic organisms absorb photons from various regions of the solar spectrum into chlorophyll molecules in thylakoids, or intracytoplasmic membranes; plants do the same in intracellular organelles called chloroplasts.

As mentioned in the first chapter, photosynthesis can be viewed as the planet breathing: it accounts for a global annual CO_2 flux of 124 PgC/yr and an annual O_2 flux of $\sim 10^{11}$ t/yr. But it can also be considered as the planet's primary nervous system – generating a basic voltage that powers the world's life.

The importance of photosynthesis to human and ecosystem sustainability is under-appreciated in global and national governance circles. Treaties and laws do exist to conserve forests and ecosystems that undertake photosynthesis, but the process itself has never been the subject of a specific piece of legislation, an international declaration or a treaty. This being the case, let's review its basic components. In photochemical conversion the 'harvested' photons' energy creates spatially separated electron/hole pairs. The holes are captured by the oxygen-evolving complex (OEC) in a protein known as photosystem II (PSII) to oxidise water (H_2O) to hydrogen and oxygen that are released to the atmosphere. At the core of this process is the tetra-nuclear manganese/calcium cluster (Mn_4CaO_5), which when characterised to the level of 1.9 angstroms has a 'distorted chair' shape. Why this particular shape should have been preserved at the heart of the natural photosynthetic process is a mystery that may have implications for design of artificial water catalysts and for engineering quantum coherence for electron transfer.

The overall photosynthetic process can be represented as free energy (ΔG) in eV $\rightarrow 2H_2O \rightarrow 2H_2 + O_2$. The electrons thereby produced are captured in chemical bonds by photosystem I (PSI) to reduce NADP (nicotinamide adenine dinucleotide phosphate) for storage in ATP (adenosine triphosphate) and NADPH (nature's form of hydrogen). In the relatively less efficient 'dark reaction' ATP and NADPH as well as carbon dioxide are used in the Calvin-Benson cycle to make food in the form of three carbon sugars, then sucrose and starch via the enzyme RuBisCO. RuBisCO is referred to by some researchers as a 'lousy' enzyme, chiefly because it requires so much of the plant's energy (almost 60 per cent) to operate. Large research projects are now sequencing thousands of different forms of the RuBisCO enzyme and comparing related kinetic data – one aim being to gain insights about how to improve upon the carbon dioxide reduction process.

In its present technologically un-enhanced form, photosynthesis globally traps 4,000 EJ/yr solar energy as biomass. This compares to a global human energy use of 500 EJ/yr. It is this stark fact, as well as the potential of photosynthesis to generate carbon-neutral hydrogen fuel (which when burnt creates fresh water) and oxygen, as well as to absorb carbon dioxide and make it into food (using the problematic enzyme RuBisCO), that highlights the primacy of its candidature for a global NES project.

The global biomass energy potential for human use from photosynthesis as it currently operates globally (focusing, for example, on crop and forestry residues, energy crops, and animal and municipal wastes) has been variously estimated to be 33 to 1,135 EJ/year, 104 EJ/year or 91 to 675 EJ/year (remembering that human energy consumption is approximately 500 EJ/yr). Yet photosynthesis has evolved in climatic conditions quite different from those we now face. Much of the light it absorbs is radiated away as heat through a variety of photoprotective mechanisms. The oxygen produced in the reaction is damaging and much energy has to go into self-repair processes. There are, in other words, many reasons why we now should try to use nanotechnology to improve the photosynthetic process.

One option for a global NES photosynthesis project involves focusing on biofuels such as biodiesel and sugar, or starch-based ethanol using synthetically modified biology (SynBio). Making SynBio biofuels can involve creation of minimal genomes upon which useful nanostructures can be built, as well as artificial cells or standardising and connecting genetic components with specific functions. Related approaches include genetically manipulating or even synthetically reproducing photosynthetic algae and bacteria to maximise their light capture and hydrogen or starch-based fuel production activities. One technique involves sulphur starving cyanobacteria so they produce hydrogen; another uses genetic modifications to make butanol.

Such SynBio options offer to create fuels with high energy content (if hydrogenated) and portability as well as compatibility with existing petroleum-based transportation infrastructure (biofuel ethanol currently has an energy density of about 25 mj/kg compared with standard petrol, which has 47 mj/kg). SynBio fuels take advantage of the existing complex cellular machinery of natural photosynthesis to capture solar energy in cell wall polymers (for instance, cellulose, hemicellulose and lignin) – this energy is then capable of being readily accessed by burning or complex bioconversion processes. SynBio transportation fuels based on lignocellulosic biomass (for example, grasses, wood and crop residues) offer a renewable, geographically distributed (growing on marginal agricultural land without significantly compromising global food crops) alternative to petroleum.

Nonetheless, SynBio fuels must still be burnt and release carbon dioxide to the atmosphere. SynBio fuels may additionally run up against the types of objection raised against genetically modified foods. Further, working on the basis that plants are less than 2 per cent efficient in converting solar energy into carbohydrates (producing say 0.5 W/m^2), then even if 3,000 m^2 per person is devoted to biofuels, unless SynBio induces major improvements, they will indirectly (via intermediate energy carriers) contribute only 36 kWh/day per person (when average global consumption is 125 kWh/day).

Nanotechnology, however, is capable of providing even more efficient, 'fully flexible', wholly artificial (non-life-based) photosynthetic systems that can be incorporated into engineered structures. Large research teams in many nations now are actively redesigning photosynthetic components, such as light capture antennae, artificial reaction centre proteins, organic polymers and inorganic catalysts, to achieve low-cost, localised, 'off-grid', direct (without intermediate energy carriers) use of sunlight to split water and achieve hydrogen for fuel cells or compression and hyper-cooling to form a liquid fuel that when burnt produces fresh water.

A global NES project focused on nanotechnology-based artificial photosynthesis (AP) offers the prospect of overcoming major limitations of natural photosynthesis. These include: 1) the light capture, water splitting and carbon dioxide reduction components need no longer be co-located geographically and temporarily; 2) the 'bioelectron flow' generated by the primary photochemical processes need not be localised within the organism; 3) significant water loss through transpiration need not accompany the CO_2 uptake process; 4) the heat radiated away from photosynthetic organisms to assist their survival can be retained and utilised; and 5) it will not be necessary for all of the biochemical-biophysical reactions to occur in the presence of damaging oxygen.

The last is a very important point. Natural systems doing photosynthesis expend a major metabolic effort in dealing with the toxic effects of reactive oxygen species. In H_2-generating organisms the H_2 production occurs only in subcellular regions from which background O_2 has been excluded. Similarly, given the amount of energy that goes into the enzyme RuBisCO, it will be an enormous advantage of a successful globalised artificial photosynthetic system that the 'light' and 'dark' phase reactions may be uncoupled both in terms of the actual energy–material flow balance and even the physical requirement to be physically co-located in space.

A global NES project not only clarifying the basic principles of artificial photosynthesis, but also advancing it into a globally marketable product in time to make an appreciable difference to problems such as climate change will involve scientific advances linking nanotechnology with fields as diverse as biology, physical and organic chemistry, physics and genomics. The scientific problems that must be solved are diverse and complex and involve three core areas: light capture, water splitting and carbon dioxide reduction.

In relation to light capture, an NES AP project could produce, for example, nanostructured materials absorbing photons from a much wider region of the solar spectrum. In antenna chlorophyll molecules in cell membrane thylakoids, for example, photon absorption currently is restricted primarily to 430–700 nm with little in the visible region between these wavelengths, though these may be covered by accessory chromophores-carotenoid polyenes, phycoerythrins and phycocyanins. A global NES AP project, for instance, could develop mesoporous thin film dye-sensitive solar cells of semiconductor nanoparticles or carbon nano-tubes allowing multiple stacking of solar cells with increasing band gap energies to exploit the solar spectrum more profitably.

New light capture techniques developed under a global NES AP project could separate charge efficiently over macroscopic distances using inexpensive materials. Nanomaterials and hybrid organic-inorganic nanostructures could improve the conversion efficiency of existing PV units. Such a project may lead to applied understanding of how the molecular dynamics of the bridge molecules in a donor-bridge-acceptor system control long-distance electron transport in photosynthetic systems. This understanding will allow the preparation and characterisation of new covalent building blocks that both initiate photo-induced charge separation and promote long-distance charge transport by self-assembling into extended non-covalent structures.

A global NES AP project may allow the excitation energies of each nanoscale arranged pigment chromophore (excitons) in the antenna array to be mapped in terms of protein spatial, orientational and energetic factors, facilitating the construction of artificial photosynthetic electron pathways to the reaction centre that perform a single quantum computation, sensing many states simultaneously and so enhancing the efficiency of energy capture and transfer at physiological temperatures.

Much of the research work under a global NES AP project is likely to involve ways to decrease the entropy in the system by imposing constraints on the spatial relations amongst the pigments, donors and acceptors. Nanotechnology-designed molecules that prolong charge separation will probably apply one of two strategies: chemical bonds in supramolecular structures, or supports such as polymers, zeolites, sol-gel glasses, lipid membranes and self-assembled films. AP devices in widespread use may involve metal complexes and molecular assemblies on surfaces and in rigid media.

Such a project could develop water splitting catalysts that stay active for extended periods of time or can be easily regenerated and made from readily available and inexpensive materials. Storing the equivalent of current energy demand will require splitting more than 10^{15} mol/year of water. If artificial photosynthesis systems could use around 10 per cent of the sunlight falling on them, they would only need to cover 0.16 per cent of the Earth's surface to satisfy global human energy consumption.

PSII in higher plants is a highly complex, membrane-based enzyme made up of 27 protein subunits and 32 cofactors involved in electron transfer and light harvesting – any synthetic mimic of this protein must be simpler and involve the following essential cofactors: a high potential photo-oxidisable chlorophyll complex, a redox-active tyrosine and a tetra-nuclear manganese/calcium cluster (Mn_4CaO_5). The most widespread AP water catalytic system will involve inexpensive and self-repairing components that operate under ambient conditions at neutral pH with non-pure water. Such water oxidation centre will also need to be stable to air, water and heat. Synthetic proteins (maquettes) developed as part of this project may allow testing of engineering principles for artificial photosystems and reaction centres.

Amongst the inorganic catalysts that could allow sunlight to split water into hydrogen for fuel cells more robustly than biomimetic materials, cobalt oxide is relatively abundant but is often obtained inequitably and is toxic in disposal. Clusters of nano-sized cobalt oxide (Co_3O_4) in parallel nanoscale channels of silica have catalysed about 1,140 oxygen molecules per second per cluster, which is commensurate with solar flux at ground level (approximately 1,000 Watts per square metre). In another technique, small angle neutron scattering analysis has showed that the protein known as light harvesting complex LHC-II, when introduced into a liquid environment that contained polymers, interacted to form nanoscale lamellar sheets similar to those found in natural photosynthetic membranes. Complete understanding of multielectron redox reactions may be achieved and new generation semiconducting oxide photoelectrodes developed.

Another component of a global NES AP project may involve multi-walled carbon nanotubes being used to redesign the PSII manganese-calcium-oxygen cluster to produce water splitting electrodes with improved efficiency, operative voltage, current density and operational stability. Manganese, after all, is the catalyst settled upon by plants after billions of years of experimentation.

A potential 'downstream' outcome of a global NES AP project could be nano solar cells that provide the hydrogen for fuel cells in zero emission cars. As one example, the company Rising Sun Technologies has obtained a licence to make nano-crystalline metal oxide films that react with photons in incident light. This solar cell splits water into hydrogen and oxygen. The hydrogen is then split in a car fuel cell into a proton and an electron, a membrane allowing the proton to pass through while the electron generates an electrical current in a copper wire. A second catalyst then combines the proton and electron with oxygen to form water. The aim is to produce cars that need no combustion of old photosynthesis fuels and so produce no emissions. The car has nanosensors to warn of hydrogen leaks, and an automatically activated ventilation system and cut-off valve. The manufacturers are planning a self-cleaning, solar storing body, one with a polmer-nanofibre shell that is stronger but lighter, and wafers around the hot parts to convert thermal to electrical energy.

An NES AP project could be promoted to the public as having a major focus on the production of cheap and abundant hydrogen. The US government decided early in the twenty-first century not to support hydrogen cars owing to: 1) the cost and durability of fuel cells wasn't adequate; 2) there was a technical inability to store large volumes of hydrogen fuel; 3) an absence of a carbon-free way of generating the hydrogen; and 4) the need to build a nationwide infrastructure. The production of these ultra-low emission vehicles will require that an NES AP project facilitates catalyst improvement in: 1) reactivity, selectivity and yield; 2) optimising and reducing active species loading levels; 3) durability and stability; 4) reducing reliance on precious-metal-based and corrosive catalysts; and 5) cost, energy intensity and environmental safety.

Another major 'downstream' part of a global NES AP project could be public transport relying on hydrogen fuel cells in nanotechnology batteries where electricity is generated by combining stored hydrogen with oxygen from the air – water vapour and heat being the only by-products. Nanoscale structures adopted from the semiconductor industry are likely to dramatically increase the catalyst surface area and replace expensive rare earth (for example platinum, iridium or ruthenium) catalysts.

A global NES project could become synonymous with a new class of visible light photocatalyst for hydrogen fuel production that can be applied like paint for houses, commercial and industrial buildings on windows, roofs and walls. Nanotechnology under a global NES AP project may convert large desalination plants into dual-use hydrogen fuel and fresh water production plants that are cheaper and cleaner to operate, reducing reliance on dams and precious river systems.

Gradually under the impetus of a global NES project in this area, a readily marketable AP design and product will emerge and costs will be lowered as inefficiencies are removed. The collaboration of global public and private sectors in funding an NES AP project will mark a major shift from a global economy predicated on bulk material single-purpose fixed processing using competitive advantages in natural resources, scale and cheap labour. Instead, the NES AP technology developed is likely to have many of the features of a natural organism, including flexibility of purpose and location, as well as capacity to not only respond to, but also positively contribute to sustainability in varying environmental conditions. Yet, as has been pointed out so often through this book, however wonderful is the science behind a global NES AP project, if it is not enabled by public policy and accepted by the general public it won't be made a reality.

9.2 GLOBAL ARTIFICIAL PHOTOSYNTHESIS (GAP) PROJECT: PROMOTING THE PUBLIC DEBATE

In 2010 I played a small role (in Professor John White's team) for the CSIRO-led academic collaboration on nanotoxicology research into cerium oxide, zinc oxide and nanosilver as part of Australia's contribution to the OECD Working Party on Manufactured Nanomaterials. At one of those meetings I met a public servant – Dr Claire Findlay of the Europe, Americas and Strategy International Science Branch, Science and Infrastructure Division Australian Federal Department of Innovation, Industry, Science and Research (DIISR).

Claire liked the idea for a Global Artificial Photosynthesis (GAP) project. 'It's a pity you didn't speak to me a little earlier, because we have an international science linkage scheme that might have funded a GAP conference for you – it's closed now,' she said. 'No problem,' I replied. 'Don't you often have some residual funds left over after such projects?' 'Yes,' she said. 'Well, what if I matched you dollar for dollar?' 'How much are you asking for?' Claire asked. 'It will still have to go through the full competitive review.' 'Of course,' I responded. I confidently plucked from the ether a figure I considered the minimum required to efficiently organise such a conference.

To obtain the remainder I then contacted in turn Mandy Thomas Pro Vice-Chancellor (Research and Graduate Studies), the Dean in the College of Law, Professor Michael Coper and Professor Aidan Byrne, Dean in the College of Medicine, Biology and the Environment. My line to them, seriatim, was: It's the first conference dedicated to the creation of a Global Artificial Photosynthesis project. It's an Australian idea. It'll look really bad if the university doesn't back it.'

It was then time to organise the GAP 1 conference. I had a meeting with senior artificial photosynthesis researchers at my university (Ron Pace, Warwick Hillier, Elmars Krausz, Jan Anderson and Fred Chow). They gave me encouragement and suggestions. I selected Lord Howe Island as it was UNESCO World Heritage listed. That meant not only was it beautiful enough (and readily accessible by connecting flights from Sydney and Brisbane) to lure researchers out of the busy international conference season, but it might facilitate UNESCO interest. Indeed, I was able to obtain endorsement from the UNESCO Natural Sciences Sector and to have the event listed as an official one for the UNESCO International Year of Chemistry. The image of Lord Howe Island I hoped might become a symbol for a nanotechnology-supported sustainable world.

How can the general public most efficiently be convinced (as they must for the success of the NES AP project) that capturing, converting and storing secure, carbon-neutral, sustainable energy, particularly in the form of fuel, from its most abundant source, the sun, is the most important scientific and technical challenge facing humanity and scientists involved in nanotechnology research?

One approach would be to argue that a Global Artificial Photosynthesis (GAP) project will draw on lessons from the well-known and successful Human Genome Project (HGP). Such a macroscience NES AP project, as has been argued, can justifiably be represented as a crucial point of intersection for the physical research that resulted in nanotechnology, morality, bioethics, domestic and international law as expressed in a global social contract. Industry investment and involvement (either as suppliers of equipment or resources or customers of outputs) in any such project will have to be encouraged, but under terms that protect the public and environmental interest. The tensions, for example, between public and private rights exhibited in the final stages of the HGP are now part of scientific folklore. Corporations investing, for example, in a global NES AP project will need to have their investments protected by patents. Yet, if the governance arrangements are not appropriate such intellectual property protections may stifle scientific research and the pace and value of its outcomes.

We are now collectively engaged in an uncontrolled global experiment with a highly restricted time frame about how best to live on Earth and ensure that the children of our children have the same opportunities as us to flourish here in an environment that nurtures their better qualities. Historians continue to document the many civilisations that have failed because they acted upon ethical and social norms that did not adequately check population growth or exploitation of natural resources. We should not be surprised that those in charge of global governance appear not to be heeding those lessons. What is more perplexing is how long it is taking human ingenuity to resolve the complexities and reveal the symmetry of science and social policy likely to resolve the current global environmental crisis.

This context, which is driving the current global policy debate about sustainability, is ripe to support a global NES AP project. In some ways the climate change problem is providing a motivation not only for a rapid transfer away from fossil fuels, but towards a global civilisation powered by AP – a new, decentralised, 'off-grid' energy system. The United Nations Earth Summit, for example, is a global project aimed to secure renewed political commitment to sustainable development, to assess progress towards internationally agreed goals on sustainable development and to address new and emerging challenges. It is backed by a resolution of the United Nations General Assembly focusing on two specific themes: a green economy in the context of poverty eradication and sustainable development, and an institutional framework for sustainable development. A global NES AP project fits cleanly within such a global governance framework and this provides a valuable potential way in which its value can be introduced to the public.

Similarly, the International Council of Science (ICSU), the International Social Science Council (ISSC) and the Belmont Forum, representing the International Group of Funding Agencies for Global Change Research (IGFA), proposed a ten-year research initiative on *Earth System Research for Global Sustainability*. This involved an extensive consultation process resulting in the identification of the *Grand Challenges for Earth System Science for Global Sustainability* integrated into a global Earth System Analysis and Prediction System (ESAPS) to assist policy-makers. A global NES AP project could be presented as readily fitting within such a governance framework.

Other governance ideas for presenting a GAP project include that it would provide targeted scientific leadership supporting policy initiatives such as the 2009 Copenhagen Accord to keeping global warming to less than 2°C above pre-industrial levels with undertakings concerning mitigation, financing (including the Copenhagen Green Climate Fund) and, in particular, establishing a technology transfer mechanism 'to accelerate technology development and transfer [...] guided by a country-driven approach'.

Another strategy for communicating to the global public the benefits of a GAP project is that it promotes governance structures to help coordinate and accelerate the activities of numerous competitively funded nanotechnology-focused AP research teams in many developed nations. A dozen European research partners formed the Solar-H2 and Eurosolar Fuels AP networks coordinated by Stenbjörn Styring from Uppsala University, Sweden, and supported by the European Union. The US Department of Energy (DOE) *Joint Center for Artificial Photosynthesis* (JCAP), led by the California Institute of Technology (Caltech) and Lawrence Berkeley National Laboratory (Nate Lewis and Peidong Yang), has US$122 m over five years to build a solar fuel system. Caltech (Harry Gray) and the Massachusetts Institute of Technology (Dan Nocera) have a $20 million National Science Foundation (NSF) grant to improve photon capture and catalyst efficiency, while several Energy Frontier Research Centers funded by the US DOE are focused on AP. Yale (Gary Brudvig) and Arizona (Devens Gust, Tom and Anna Moore) Universities also have major AP projects. Important AP work is also being performed by Bertil Andersson's group at Nanyang University in Singapore and Negishi's group based at Hokkaido University in Japan, and James Durrant, Bill Rutherford and Jim Barber at Imperial College London. In Australia significant AP groups exist at the Australian National University (Ron Pace, Elmars Krausz and Warwick Hillier), Melbourne University (Paul Mulvaney), Woollongong University (Gordon Wallace and David Officer) and Monash University (Leone Spiccia and Doug MacFarlane).

A GAP project must overcome distinct organisational as well as scientific challenges. The scientific challenge for the Human Genome Project (HGP) was more clearly defined in basic research than is the case with GAP (which has a strong marketable product focus). The HGP (1990–2003) was an international collaborative scientific research project with a primary technical goal: sequencing of chemical base pairs that make up DNA to identify and map the approximately 20,000–25,000 genes of the human genome from both a physical and functional standpoint. Starting with an idea for international collaboration, the HGP grew into a consortium where government-sponsored sequencing was performed in universities and research centres from the United States (National Institutes of Health), the United Kingdom, Japan, France, Germany, China, India, Canada and New Zealand. A parallel project was conducted by the private industry Celera Corporation. In 2001, in a public relations triumph, the human genome project international consortium announced the publication of a draft sequence and analysis of the human genome.

GAP project work, however, will not be centrally focused on fully resolving the mechanism of natural photosynthesis. It will be oriented to producing a viable commercial product. Like the HGP, the GAP project is likely to be distributed across a variety of laboratories in different nations, rather than being focused in one place like other macroscience projects of public notoriety such as the European Organization for Nuclear Research (CERN) or the international project on fusion energy (ITER). Having central showcase laboratories could assist a GAP project based on lessons from CERN's scientific popularity and public profile, which have been enhanced by allowing many nations to fund new equipment (such as the Large Hadron Collider) open to use by independently funded physicists from around the world. A GAP project might also benefit in terms of public profile from incorporating governance mechanisms such as that at the Hubble Space Telescope (funded by NASA in collaboration with the European Space Agency), which allows any qualified scientist to submit a research proposal, successful applicants having a year after observation before their data is released to the entire scientific community.

The International Society of Photosynthesis Research (ISPR) is likely to play a strong scientific and media coordinating role in the creation and advancement of any GAP project. The ISPR coordinates a triennial international congress on photosynthesis, as well as being required by its constitution to 'encourage the growth and to promote the development of photosynthesis as a pure and applied science'. It is also required to 'promote international cooperation in photosynthesis research and education'. ISPR membership spans six continents and our members work across academia, education and training, as well as in government, industrial and commercial research environments; so ISPR plays a key role in uniting the photosynthesis research community internationally.

Another international institution whose association with a GAP project is likely to raise the latter's public profile is the International Renewable Energy Agency (IRENA), which involves 147 countries and the European Union. About 30 member states have now ratified the IRENA Treaty. The American Council on Renewable Energy (ACORE) is the US affiliate of the World Council for Renewable Energy (WCRE), and is dedicated to bringing renewable energy into the mainstream of the US economy and lifestyle through information and communications. ACORE provides a common platform for the wide range of interests in the renewable energy community including technology and service companies, associations, utilities, end users, professional service firms, financial institutions and government agencies. ACORE serves as a forum through which these parties work together on common interests.

Other organisations likely to be important to the governance processes, scientific collaborations and public profile of a GAP project include EUROSOLAR, the non-profit European Association for Renewable

Energy, which includes political and scientific institutions, companies and associations as well as parliamentarians, scientists, architects, engineers, craftsmen, farmers, teachers and citizens, engaged in and dedicated to the cause of renewable energy expansion. The EUROSOLAR charter in developing and stimulating political and economic strategies, concepts and legal frameworks for renewable energy suggests it may play a coordinating regional role in any GAP project. The European Union's Renewable Energy Directive (RED) and related National Renewable Energy Action Plans (NREAPs) could be other areas of governance interest for a GAP project. The United Nations, UNESCO and World Health Organization (WHO) may also play a role. UNESCO could be particularly important if it initiated work towards a scoping declaration enunciating the governance principles (in terms of bioethics and international human rights) that should govern the development of artificial photosynthesis. Chief amongst these might be the statement that photosynthesis in its natural form should be placed within the global public goods category of 'common heritage of humanity'.

9.3 NES GAP PROJECT: GOVERNANCE MODELS AND POLICY ISSUES

The GAP 1 conference at Lord Howe Island by all accounts was a great success. The winds were not too strong, which meant that all the planes could land on the short airstrip between two beaches and below the island's two mountains. My favourite memories included Peidong Yang, head of part of the Joint Center on Artificial Photosynthesis (JCAP) at the University of California Berkeley, and Dan Nocera from MIT with their arms round each other's shoulders. Professor David Tiede from the Argonne National Laboratory missed his connecting flight in the US, but we managed to get him on a twin engine plane owned by the owners of the cafe that was catering the morning teas for the conference.

The sessions began at 8 a.m. and concluded at 10 p.m. with 12.30 to 4 p.m. as free time to relax, discuss and wander about the island. Speakers had arrived from almost all the major AP projects in the world as well as leaders in photovoltaics, hydrogen production and storage and quantum coherence in electron transfer plus GAP-related governance issues. At the end of each conference session high school students from James Ruse (NSW), Geelong Grammar (Vic), Radford College (ACT), Canberra Grammar (ACT) and Narrabundah College (ACT) gave brief presentations involving thought-experiments about the scientific or governance aspects of a GAP-powered world. The researchers were asked to bring a book that would inspire a young person to get into science and these were autographed by all participants and presented to the students during the conference dinner. Brian Greene's *Fabric of the Cosmos*, Matt Ridley's *Rational Optimist* and Oliver Morton's *Eating the Sun* were popular. Some researchers brought bound copies of their PhD!

In the evening, we would end the day by discussing the models and policy issues involved in setting up a GAP project. It is to these that we now turn here. Let's start with some of the most obvious legal issues likely to be faced. Many of the nanotechnological techniques and structures, as well as the artificial proteins involved in a GAP project will be the subject of patent or other intellectual monopoly privilege (IMP) claims. In general terms, this will require that their inventors claim such contributions to be novel, inventive (non-obvious in the USA) and useful, with a specification complete enough to allow others to make the device without undue experimentation. University legal departments, unless restrained, may well swamp research seeking to collaborate under a GAP project with a plethora of confidentiality agreements protecting institutional IMPs and seeking to gain a percentage of royalties on any marketed outcome.

The process of photosynthesis being as central to life on Earth as DNA, there are likely to be major public debates over whether patents should be allowed over any part of it. Such debates will be unlikely to inhibit some patents being taken out over aspects of AP. The US Supreme Court and courts in many other nations, for example, have ruled that, although the human genome as a whole is not patentable, individual genes can be patented if they are isolated and purified.

GAP research and development will also face major issues about whether patents should cover AP products as well as AP processes and functions. It is likely that in the US the 'utility' for an AP patent (as is the case for DNA) will be that it must be specific, substantial and credible. If GAP IMP ownership becomes fragmented, researchers in the field will find follow-on research hampered by the high cost and difficulty in negotiating contracts with large numbers of IMP owners. Each individual GAP patent owner, for example, without some prior licensing and sharing arrangement, will have little incentive to share information, or work together to resolve problems when priority may be in doubt.

Appropriately regulating industry involvement (either as suppliers of equipment or resources or customers of outputs) in a GAP project will be another major governance issue given the tensions between public and private rights exhibited in the final stages of the HGP. Lessons from the SEMATECH (SEmiconductor MAnufacturing TECHnology) non-profit consortium may be that while large-scale public sector funding and industry investment are necessary for initial momentum, global impact with a marketable product requires division into pure research and manufacturing subsidiaries.

The Center for Revolutionary Solar Photoconversion (CRSP) could provide a useful model here. It involves public funding from two separate sources (US DOE and NSF) with supranational corporate members (including DuPont, General Motors, Konarka, Lockheed Martin, Sharp and Toyota).

Chemistry will assume a special role in a GAP project, because new materials must be created for solar capture and conversion and because new catalysts are needed for the desired chemical bond conversions. Other areas of GAP research will include physics, quantum physics, imaging, cell biology, regulation and policy. Governance scholars will seek to create processes that not only facilitate interdisciplinary research, but also minimise institutional rivalry and engage in public debate concerning GAP ethics, resource allocation and national energy security policies.

An open-access GAP governance model (perhaps in a UNESCO declaration or convention on the *Bioethics and Human Rights of Natural and Artificial Photosynthesis*) could emphasise funding rules requiring public good licensing, technology transfer, ethical and social implications research as well as rapid and free access to data. A public–private partnership model might involve members' access to non-exclusive licences over intellectual property. Many of the debates that will impact on GAP here are already being played out in relation to synthetic biology, and the governance solutions developed here could be good models for a GAP project. Without such attention to developing such governance arrangements, IMPs claimed over GAP components (such as antenna systems, reaction centres and water catalysts) may be hard to identify, fragmented across many owners and sometimes overly broad, making it harder for would-be GAP innovators to progress towards a globally marketable product.

As GAP research progresses towards the production and marketing phase, the need for common standards may promote a 'tipping dynamic' in which one solution owned by a single corporation quickly comes to dominate the field. GAP will probably benefit not from following the patent 'wars' of the pharmaceutical development, but from examples such as the mobile telephone industry where no single manufacturer owns every patent that covers its product-forcing sharing arrangements. The number of licensing transactions that GAP firms face could be reduced by making the standard parts of AP unpatentable. Courts are unlikely to do this, though, ostensibly because judges in specialist patent courts favour a pro-incentive-to-innovate policy over one favouring public goods.

Patent pools between the public–private sectors could be used to promote GAP collaboration, but governance structures should not violate anti-monopoly laws by excluding competitors. The complex technologies involved in AP may cause researchers competing in this rapidly advancing area to unintentionally infringe IMPs, especially when patents are allowed to be broad and numerous or vaguely written. Patent 'trolls' may infest the AP area as they have in that of biotechnology. Such firms will acquire patents (for example from bankrupt AP firms) simply to parasitically profit from the need of AP researchers and their firms. In the software industry the Open Innovation Network gets around this by buying up Linux-related patents.

Excessive AP patents may also mean that GAP firms and their researchers will tend to become locked into researching the AP parts that have become familiar, have ready legal access, are already widely used across the industry and are less likely to be the subject of IMP challenges. On the other hand, if the most widely used GAP components are subject to some form of open-access permission, then the research will proceed faster. This effect can be enhanced if GAP researchers deliberately choose to incorporate 'open' rather than closed (patented) parts. Incentives should also be created for GAP-involved firms and universities (for reasons such as raising profile, obtaining reciprocal access and building a user base) to donate part of their data to open source projects.

Additionally, companies involved in linkage projects with GAP researchers could be required to specify as part of their grant application how much IMP protection they will need. The competition for public grants would then provide a powerful incentive for limiting patent duration. Cambia's Bioforge initiative BiOS (Biological Open Source), for example, is a legally enforceable framework to enable the sharing of the capability to use patented and non-patented technology, which may include materials and methods, within a dynamically expanding group of those who all agree to the same principles of responsible sharing, a 'protected commons'. Those who join a BiOS 'concordance' agree not to assert IP rights against each other's use of the technology to do research, or to develop products either for profit or for public good.

Likewise, the Initiative for Open Innovation (IOI) is a new global facility funded through grants from the Bill & Melinda Gates Foundation and the Lemelson Foundation. The BioBricks Foundation (BBF) is a not-for-profit organisation founded by engineers and scientists from MIT, Harvard and UCSF with significant experience in both non-profit and commercial biotechnology research. BBF encourages the development and responsible use of technologies based on BioBrick™ standard DNA parts that encode basic biological functions.

Members who joined an open-source GAP collaboration might gain brief periods of exclusive ownership (much less than 20 year patent terms) in return for a promise to afterwards share data and receive access to a confidential database governed by trade secrets and copyright laws that are less expensive or restrictive than patents. Members could publish information they supplied at any time – so blocking third parties from obtaining patents (as happens with the Merck Gene Index) but promoting exchange of data.

As mentioned, a major component of an NES GAP project could be a UNESCO *Universal Declaration on the Bioethics and Human Rights of Artificial and Natural Photosynthesis*. This might provide in Article 1 (in a minimal legal obligation 'soft-law' approach):

> Photosynthesis underlies the fundamental unity of all life on Earth, as well as the recognition of its inherent dignity and diversity. In a symbolic sense, it is the heritage of humanity – a commons to be preserved for future generations and protected from corporate or national appropriation.

A GAP UNESCO *Universal Declaration on the Bioethics and Human Rights of Artificial and Natural Photosynthesis* might be a good place to start the governance process of protecting natural and artificial photosynthesis within the international law concept of common heritage of humanity. It would link photosynthesis with the human genome (given symbolic common heritage status in the UNESCO *Universal Declaration on the Human Genome and Human Rights*) as well as the moon, outer space, the deep sea bed and world natural heritage sites (for which United Nations Conventions accord common heritage status as a binding obligation under international law).

Five core components have been associated with the 'common heritage of humanity' concept that could apply to global artificial photosynthesis. First, there can be no private or public appropriation; no one should legally own photosynthesis in either its natural or artificial forms. Second, representatives from all nations must manage artificial photosynthetic resources on behalf of all; this practically necessitating a special agency (such as the United Nations WEO) to coordinate shared management.

Third, all nations must actively share with each other the benefits acquired from exploitation of artificial photosynthesis, this requiring restraint on the profit-making activities of private corporate entities and linking the concept to that of global public good.

Fourth, there can be no military utilisation of artificial photosynthesis (for example withholding access to GAP technology as a means of attacking another nation's energy security).

Fifth, artificial photosynthesis should be preserved for the benefit of future generations, and to avoid a 'tragedy of the commons' scenario. This terminology reflects Garrett Hardin's conception of 'lifeboat ethics' – the physical inability in a crisis to save all and the need to prioritise survivors. Hardin considered that the creation of any common good through communal ownership in a limited resource (such as equitable access GAP technology) invites its destruction by free-riders seeking to exploit that universal access. Such a view is countered, however, by the proven capacity of humanity (when given the freedom to do so) to develop robust processes for sharing and by the use of the practically unlimited (on normal timescales) resource of sunlight in GAP products.

9.4 REFINING THE PRINCIPLES OF PLANETARY NANOMEDICINE

Imagine that in 2284 you travel to the rainforests of Venezuela to visit the Yanomami tribe. The Yanomami are still one of the most isolated indigenous groups on the planet and inhabit the 8 million hectare Alto Orinoco–Casiquiare Biosphere Reserve (created in 1992). The tribe also roam amongst 10 million acres in Brazil.

You ride in your solar-hydrogen-powered car from the lowland in the plateaux of Casiquiare Canal, along roads (made with photosynthetic materials) that pass through tropical rainforest. You head to the high Cerro Marahuaca in the northeast and the Amazonian *caatinga* country. Dominated by little thorny plants, the land here experiences only two seasons: winter, now, when it's very hot and dry, and the summer of November–December when it's hot and rainy. You see trees protruding their roots amongst the yellow-grey soil and stones to absorb water before it evaporates and shedding their leaves to reduce transpiration.

In the past now are confrontations between the Yamomami and miners or loggers who'd been illegally entering and working on their land. The government with the assistance of international agencies has developed a model for sustainable management in the Upper Orinoco region. They have organised community stores, a library, public transportation, recreational facilities and agricultural improvements all incorporating artificial photosynthesis nanotechnology for energy, food and clean water.

You walk into a village of about 200 people sleeping under a common roof made of leaves, vines and tree trunks called the *shabono*. *Shabonos* have a characteristic oval shape with open grounds in the centre measuring an average of 100 yards. The *shabono* itself is the perimeter of the village, if it has not been fortified with palisade walls.

In the middle of the roof of the *shabono* is a GAP solar fuel and food unit provided for the village by a charitable foundation. It uses a nanotechnological assembly of porphyrins and fullerenes to accelerate photo-induced electron transfer that splits water to make hydrogen fuel and then absorbs carbon dioxide to make basic food and fertiliser components. It is highly efficient and absorbs light from a wide range of the solar spectrum, transferring the resulting excitation energy to a reaction centre where photochemistry occurs.

As a result of incorporating GAP nanotechnology in their community the Yanomamö, as they call themselves, are no longer dependent upon the forest. Now they enter it to supplement their diet with bananas, gathered fruit, hunted animals and fish, as well as for spiritual refreshment. Many people regularly visit from across the world to briefly share the Yanomamö experience of living in harmony with nature.

One illustrative outcome of fully implemented global artificial photosynthesis is that forests will no longer need to be instrumentalised as sources of fuel or as agricultural, residential and industrial land as well as carbon sinks. They now can be managed more sustainably as a 'rented' service from nature as national parks, sources of quality timber and alleviation of poverty amongst the rural poor, conserving and restoring biodiversity, protecting habitat and soil and water resources.

If this positive vision does not eventuate, but our biosphere collapses, as many scientists predict it may within the next hundred years or so, most business and policy leaders will regard such a tragic outcome as not the consequence of particularly amoral or arrogantly self-destructive rejections of voluntary or imposed constraints on freedom of will by any individual human or collective. Indeed, those involved in shaping the global governance of the industrial technologies contributing, for example, to anthropogenic climate change, weapons of mass destruction, as well as inequitable access to essential medicines, or basic food or water supplies, would probably be genuinely perplexed when told the laws and policies they've successfully lobbied for are creating such harms through irreconcilable anomalies with respect to established social virtues.

The enlightenment philosopher Immanuel Kant might have ridiculed a vision like that expressed above with humans eventually living sustainably on Earth in harmony with their ecosystems. Kant in his writings often criticised our yearning to return to a poetic golden age in which we are content with the bare necessities of nature and there is complete equality and perpetual peace among men. He wrote that humans cannot go back to such a state because not only does it not exist, but if it did it would not satisfy us. Strangely, however, such a conclusion seems inconsistent with Kant's own postulates that reality consists of many elements (such as time and space, as well as commitment to apply universal ethical principles) that don't fully correlate with our common sense perceptions.

Since the eighteenth century, humanist writers particularly of the Utopian tradition have tried to suggest ways in which all human beings may be able to live harmoniously together in a way that is meaningful. Many see the bioethics and international human rights movements as offshoots of such enlightenment thinking. The fact that a substantial proportion of humanity's most influential and admired leaders have and continue to not only yearn for, but strive to build, such a golden age, suggests that hypothesis is derived from a deep pattern of reality. It is a basic postulate of science-based natural law (as advocated here) that this type of idealistic social and environmental vision acknowledges laws every bit as real as those of modern physics that also baffle common sense.

Alternate visions of a GAP-fuelled future (or reality in a parallel universe) might involve organic zones from which nanotechnology is excluded, a bit like national parks today, or fortresses in which nanotechnology sustains and protects prosperous populations from less privileged and increasingly hostile enemies. Such visions, however, as has been explained here, are not coherent with the universally applicable moral principles developed from science-based natural law and social contract thinking.

The fact that the universe permits an infinity of options doesn't endow each with equivalent moral status. While nanotechnology, for instance, may allow local production of energy, food, clean water and fertiliser and while domestic nanofactories can provide consumer objects, those innovations will not of themselves lead to a more sustainable and meaningful world.

This book has analysed the type of social virtue that should arise in a world whose human activities are primarily powered by the outputs from an NES artificial photosynthesis project. Without the element of optimism involved in imaging a global society that places the collective virtue of respect for environmental sustainability alongside respect for human dignity at the heart of its social contract, it will be harder to produce leaders willing to accept responsibility for making it happen. Surely it is worthwhile exploring the hypothesis that striving to implement governance structures and principles that facilitate a more environmentally sustainable, just and equitable world through the use of GAP technology is coherent with fundamental patterns of physical symmetry.

This book has argued that both nanotechnology and governance involve construction from very small components. The case has been made that, just as in materials science, perfect resolution of the smallest scales (according to Heisenberg's uncertainty principle) requires a probe with infinite energy, discovering the fundamentals of governance norms entails an eternal perspective.

The task of proving that 'laws' (both physical and moral) go 'all the way down' in reality is complex because as matter and time reach their smallest extent (the Planck scale) quantum effects make everything including points, velocity, length and curvature fluctuate, and become indeterminate and uncertain as soon as we try to observe them. Thus, at a Planck length of 1.6161×10^{-35} (which squared gives the Planck area of 2.61177×10^{-70} m^2), or a Planck time of 5.39072×10^{-44} s, or Planck mass of 2.17665×10^{-8} kg (10^{-19} times the mass of a single proton) gravity starts to become as strong as the electromagnetic and strong (builds protons and neutrons out of quarks and gluons) and weak (links different 'flavours' of quarks and leptons) nuclear forces distort the very notions of space and time. In fact, the Planck length is 100 billion billion times smaller than anything that has been explored experimentally. When we finally manage to peer at the patterns of energy emerging at this level, might we end up observing some fundamental aspect of ourselves – perhaps even conscience in its purest form?

Planck units, like the elementary electric charge and speed of light, are natural 'laws' in that they are based on universal physical constants, not dependent on any human definition. As we've seen, the geometer Shing-Tung Yau considers that geometry might be the ultimate source of such 'natural laws', shaping multiple dimensions of space and time even in the absence of matter. The string theorist Edward Witten has argued for a new type of quantum geometry in which all the basic 'laws' of matter might be harmonics of strings operating in those multiple additional dimensions.

Putting the case another way, in considering our moral obligations as a species towards nanotechnology, this book has posited that the fundamental 'laws' of global governance systems are not merely contingent manifestations of political compromise and judicial interpretation, but (when considered from the appropriate perspective) likewise aspects of this basic geometric and mathematical symmetry of the universe. This, it has been argued, is the philosophical perspective out of which a successful global NES GAP project will emerge.

Nanotechnology thus does not represent the end of a frontier, but rather the beginning of a particular form of exploration about how our world is constructed. In the rush to commercialise nanotechnology this perspective is often lost sight of. Unfortunately, we are running out of time to conduct more social experiments about which model of nanotechnology development is the best one. Population growth, industrialisation-driven climate change, nuclear and biologic weapons in the hands of terrorist organisations and states, degradation of the biosphere – these great global public health problems create challenges that on any rational analysis must be solved immediately if we are to survive. Yet efforts to push nanotechnology research in this direction have been unsystematic and targeted more at national and corporate interest than global benefit.

Nanotechnology is often described as an 'enabling technology'. This should mean that it has the capacity to enhance the scope and application of existing approaches to problems faced by our species. Yet, as we've discussed, it is rare for that technology to be focused as part of a systematic plan to shape the Earth into the type of place that represents a high-quality, sustainable environment.

It would be too simplistic to say that any failure of global governance structures to dedicate nanotechnology to environmental sustainability derives from an inescapable defect in the human personality. From the science-based natural law perspective developed here, that would be like saying the anode is an inescapable defect of the cathode, or that the negative is an inescapable defect of a positive electromagnetic pole.

Accepting the necessity in governance arrangements for conservative and progressive perspectives, a more important issue is how they can meld to provide our species with a shared vision of not just where we should be headed, but how, and with what values and virtues? Our ostensible leaders' apparent reluctance to make any concerted or coordinated effort to do so creates opportunities for citizens to provide such a vision more democratically through the Internet and other modes of instantaneous mass communication.

The tradition of imaging an ideal civilisation, of positing some social contract at the heart of society, has been useful in terms of generating ideas about justice and equality and respect for international human rights. Yet such thought-experiments are futile exercises if they remain essentially backward-looking enterprises without the capacity to transform to meet immediate challenges. City planners are constantly imaging population densities in relation to urban space and amenities. Why shouldn't that process be implemented with broader focus and greater particularity on a global scale? We have discussed some of the obstacles to such a vision, ranging from the dominance of supranational corporations, their intellectual monopoly privileges and use of trade law to lobby governments, as well as their corruption and misuse of public funds.

Surely one focal point for promoting this vision could be a UNESCO Declaration or Convention in which the nations of the world agree to start the process of imaging how nanotechnology could assist in producing the ideal type of society each state wishes to be able to provide its citizens. Another governance vehicle for such developments might be the treaty establishing a United Nations World Environment Organization (WEO). In just such a debate the governance arrangements of a global NES GAP project may become firmly established.

EF Schumacher, an unusually enlightened economist, argued that we cannot consider the problem of technological production solved if it requires that we recklessly erode our finite natural capital and deprive future generations of its benefits. Likewise, Buckminster Fuller warned against continuing to exhaust in a relative split second of terrestrial history the orderly energy savings of billions of years' photosynthetic energy conservation, which we should cherish as placed aboard Spaceship Earth for use only in critical, self-starter, life-regeneration situations.

The governance model of a free-market consumerist society with supra-national corporate-run production in quest of endlessly increasing economic growth and profit, far from being the apotheosis of social interaction at some mythical end of history, has demonstrated its incoherence with basic social virtues such as environmental sustainability and respect for human dignity. It is surely time to allow our ingenuity as manifested in our technology to help us move towards governance arrangements in which social progress is measured chiefly by sustainable gross domestic product, human flourishing and ecosystem sustainability.

It is a natural part of the human condition that we must exercise constant vigilance to ensure vices such as greed and selfishness are not encouraged by our legal structures, policy and education systems to the detriment of core components of our global social contract. It is our conscience that, when properly trained, encourages us to expand our sympathies and gradually our understanding of what we are fundamentally.

For a young person today, rather than asking what job you'd love to do and why, my advice is to ask: 'What am I meant to do?' or 'What is my duty?' Then give 100 per cent to everything fate and reason sends your way – review the results when you've done that, building systematically on what you do best. Soon you'll know not only in what ways you're naturally gifted, but how you can grow by completing a unique task for greater good. It's mysterious, but reliable and conformable to modern physics – we're simultaneously waves (what is my duty?) and particles (what do I want?).

Science-based natural law in this way provides a practical guide to action flowing from a hypothesised unification of quantum mechanics, general relativity and the moral law within us. It asks we posit that, though normally existing as a discreet point in time and space, the more we live for conscience and apply universally applicable ethical principles in the face of obstacles, we are also a wave linking us to others and their interests. Science-based natural law is unlikely to present this as an ideology to be believed, but rather as a hypothesis to be tested. Live, for example, a month or so according to the idea that there is something you are meant to do and the postulate is that you are likely to experience a greater number of synchronous coincidences where the universe seems to be conspiring to assist you. Then live for an equivalent period as if your moral characteristics are nothing but a manifestation of how the particles in your body are arranged. Compare the difference.

The physicist Richard Feynman once claimed that the statement with the richest information about the physical world would be: 'that all things are made of atoms – little particles that move around in perpetual motion, attracting one another when they are a little distance apart, but repelling upon being squeezed into one another'. Imagine that the words 'atom' and 'norm' were interchanged here. Thus we could say that 'human relationships and freedom to act are composed of norms – the basic constituents of principles, rules and laws in perpetual motion, attracting one another when a little distance apart, but repelling upon being squeezed into one another'. Embracing such a perspective we can say, following Thomas Kuhn in his *Structure of Scientific Revolutions*, that we can begin to plot an NES GAP project using mathematically formulated laws of physics and governance, supported by images or representations of them and exemplars – past examples used as patterns for our present calculations.

The discussions here of the place of nanotechnology in science-based natural law and in a global social contract are not merely a facile exercise in argument by analogy. Their practical implications in this context relate to how humanity should view the process of discovering the optimal form of legal regulation for the world over which it currently has unprecedented and potentially catastrophic control. They also have consequences for thinking about how and why we should now, at this critical juncture in our history, research and develop emerging technologies (such as nanotechnology) that if properly used may assist humanity to live sustainably amongst the ecosystems of Earth.

To arouse the conscience of people around the world to confront potentially catastrophic problems, some authors of fiction place sympathetic characters in situations of appropriately confronting contradiction. George Orwell did this in *Nineteen Eighty-Four* in relation to technology and totalitarianism and its disastrous impact on individual liberties. Cormac McCarthy deployed a similar thought-experiment about a destroyed biosphere in his apocalyptic novel *The Road.* One problem with this essentially negative approach, however, is that it doesn't always suggest a brighter alternative, or the so-called 'nuts and bolts' of how we should move towards it.

In taking a more optimistic view, this book might be criticised for attempting to propound some particular ideal global commonwealth in which humanity's looming problems of energy, food, water and shelter have been resolved by a nanotechnological fix. I find such 'technological fix' criticism of nanotechnology's role in public health and environmental protection (and I've heard it often) facile, unnecessary and unhelpful. Technological fixes are what have characterised the history of human social development and behaviour change.

In any event, rather than presenting a particular goal, this text is designed to present a positive and idealistic vision of the fundamental concepts underpinning nanoscience and legal regulation combining to assist humanity in creating a sustainable, coherent and fulfilling world. It urges those in nanoscience and nanotechnology, as well as policy-makers, to forgo their customary preoccupation with thinking small and short term. It asks them to enter a debate about what principles we should use to coordinate global efforts that allow a wonderful new technology to help resolve some of our major public health and environmental challenges.

One of the great benefits of an NES GAP project in this context is likely to be how it encourages global human society to embrace the preservation for future generations of non-renewable resources and break the mass media-driven dominance of our governance arrangements and even our visions and hopes by supranational corporate enterprises motivated chiefly by the need to maximise profit. People whose lives are assisted by GAP local nanotechnology-assisted production of energy, food and water will naturally moderate unnecessary demands upon them and take greater pride and satisfaction from their quality. An NES GAP project will encourage people to stop fearing nanotechnology and embrace it as likely to promote (when properly governed) high quality life amongst this and future generations. It will make feasible a world in which people are not cut off from the daily rituals of the natural environment and the other forms of life with which we have evolved and whose present perilous existence and suffering should prompt us to enhanced empathy and altruism.

That being said, any one of a number of cataclysms could overwhelm the human species and its ecosystems before an NES GAP project has had time to achieve its outcomes. These crises may create more pressing survival priorities in disjointed and precariously existing groups of people. Humanity may survive (as it has in the past) but in small pockets without the benefit of its current technology. The moral opportunity to develop nanotechnology to focus on global artificial photosynthesis may never recur.

Earth from a science-based natural law perspective appears to be a type of proving ground for human consciousness. According to this hypothesis, waves of meaning roll through this world that are simultaneously collapsing to points of moral decision in distinct temporal and spatial contexts that somehow always offer the character-transforming prospect of merging individual interests in a more universal good. The symmetric patterns of energy that gave rise to photosynthesis, to life and to us, according to such an approach, also are shaping the technology as well as the ethical and legal norms, by which humanity eventually will shape a sustainable existence on this planet. Photosynthesis may be but a component in the process whereby the light of suns gives rise not merely to consciousness, but in its pure form to conscience. Centuries from now, with that task completed, we may be finally qualified to explore the universe not just in the dimensions we know now, but also in many others that exist despite their lack of apparent correlation with our common sense and sensory experience.

Bibliography

Abbott, KW, GE Marchant et al. (2006), 'A framework convention for nanotechnology?' *Env Law R*, **36**, 10931–42.

Action Group on Erosion, Technology and Concentration (ETC Group) *Down on the Farm: The Impact of Nano-Scale Technologies on Food and Agriculture*, Ottawa: ETC Group.

Albrecht, MA et al. (2006), 'Green chemistry and the health implications of nanoparticles', *Green Chemistry*, **8**, 417–32.

Allen, MR et al. (2009), 'Warming caused by cumulative carbon emissions towards the trillionth tonne', *Nature*, **458**, 163–6.

Altmann, J (2004), 'Military uses of nanotechnology: perspectives and concerns', *Security Dialogue*, **35** (1), 61–79.

Altmann, J and M Gubrud (2004), 'Anticipating military nanotechnology', *IEEE Technology and Society*, Winter, 33–40.

Arico, AS, P Bruce et al. (2005), 'Nanostructured materials for advanced energy conversion and storage devices', *Nature Materials*, **4**, 366–77.

Arthur, WB (2009), *The Nature of Technology: What It Is and How It Evolves*, London: Penguin.

Ayala, FJ (1987), 'The biological roots of morality', *Biology and Philosophy*, **2**, 235–52.

Bala, G (2009), 'Problems with geoengineering schemes to combat climate change', *Current Science*, **96** (1), 41–8.

Balbus, JM, K Florini et al. (2007), 'Protecting workers and the environment: an environmental NGO's perspective on nanotechnology', *Journal of Nanoparticle Research*, **9**, 11–22.

Ball, P (2004), *Critical Mass: How One Thing Leads to Another*, London: Arrow.

Baslar, K (1998), *The Concept of the Common Heritage of Mankind in International Law*, The Hague: Martinus Nijhoff.

Beerling, D (2007), *The Emerald Plant: How Plants Changed Earth's History*, Oxford: Oxford University Press.

Behrenfeld, M et al. (1996), 'Confirmations of iron limitation of phytoplankton photosynthessis in the Equatorial Pacific Ocean', *Nature*, **383**, 508–11.

Bogunia-Kubik, K and M Sugisaka (2002), 'From molecular biology to nanotechnology and nanomedicine', *BioSystems*, **8**, 123–38.

Borm, PJ (2002), 'Particle toxicology: from coal mining to nanotechnology', *Inhal Toxicol*, **14**, 311–24.

Borm, PJ et al. (2006), 'The potential risks of nanomaterials: a review carried out for ECETOC', *Particle Fibre Toxicol*, **3**, 11–18 (online 12 November 2007).

Bossi, P et al. (2006), 'Bioterrorism: management of major biological agents', *Cellular and Molecular Life Sciences*, **63** (19), 2196–212.

Boulding, KE (1992), *Towards a New Economics: Critical Essays on Ecology, Distribution and Other Themes*, Aldershot, UK and Brookfield, VT, USA: Edward Elgar.

Bradford, T (2006), *Solar Revolution: The Economic Transformation of the Global Energy Industry*, London: MIT Press.

Brand, S (2010), *Whole Earth Discipline: Why Dense Cities, Nuclear Power, Genetically Modified Crops, Restored Wildlands, Radical Science and Geoengineering are Essential*, London: Atlantic.

Brown, DE (1991), *Human Universals*, New York: McGraw-Hill.

Brown, S, IR Swingland et al. (2002), 'Changes in the use and management of forests for abating carbon emissions: issues and challenges under the Kyoto Protocol', *Phil Trans R Soc London*, **360**, 1593–605.

Brown, WE (1984), 'The planetary trust: conservation and intergenerational equity', *Ecology Law Quarterly*, **11**, 495–503.

Brown, WE (1989), *In Fairness to Future Generations: International Law Common Patrimony and Intergenerational Equity*, New York: Transnational Publishers.

Buckley, B (ed.) (2003), *The WTO and the Doha Round: The Changing Face of World Trade*, The Hague; London; New York: Kluwer Law International.

Burgman, I (1960), *The Seventh Seal*, New York: Touchstone.

Cane, P (2002), *Responsibility in Law and Morality*, Oxford: Hart Publishing.

Carroll, JW (1994), *Laws of Nature*, Cambridge: Cambridge University Press.

Cathcart, B (2005), *The Fly in the Cathedral: How a Small Group of Cambridge Scientists Won the Race to Split the Atom*, London: Penguin.

Cedervall, T, I Lynch et al. (2007), 'Understanding the nanoparticle-protein corona using methods to quantify exchange rates and affinities for proteins for nanoparticles', *Proc. National Acad. SCi*, **104** (7), 2050–55.

Cello, J, AV Paul and E Wimmer (2002), 'Chemical synthesis of poliovirus cDNA: generation of infectious virus in the absence of natural template', *Science*, **297**, 1016–18.

Chaudhry, Q, M Scotter et al. (2008), 'Applications and implications of nanotechnologies for the food sector', *Food Additives and Contaminants*, **25** (3), 241–58.

Chen, I (2006), 'Thinking big about global health', *Cell*, **124** (4), 661–3.

Chen, X and HJ Schluesener (2008), 'Nanosilver: a nanoproduct in medical application', *Toxicology Letters*, **176** (1), 1–12.

Chirac, P and E Torreele (2006), 'Global framework on essential health R&D', *Lancet*, **367** (9522), 1560–61.

Chomsky, N (1999), *Profit Over People: Neoliberalism and Global Order*, New York: Seven Stories Press.

Clapp, J and D Fuchs (eds) (2009), *Corporate Power in Global Agrifood Governance*, London: MIT Press.

Cockell, CS (2007), *Space on Earth: Saving Our World by Seeking Others*, London: Macmillan.

Cook-Deegan, R (1989), 'The Alta Summit', *Genomics*, **5** (3), 661–3.

Cooper, M (2008), *Life as Surplus: Biotechnology and Capitalism in the Neoliberal Era*, Seattle: University of Washington Press.

Cross, FB (1996), 'Paradoxical perils of the precautionary principle', *Washington and Lee Law Review*, **53**, 851–925.

Crutzen, PJ (2006), 'Albedo enhancement by startospheric sulfur injections: a contribution to resolve a policy dilemma', *Climatic Change*, **77**, 211–19.

Curry, P (2009), *Ecological Ethics: An Introduction*, Cambridge: Polity Press.

Dartnell, L (2004), *Life in the Universe*, Oxford: One World.

Davi Kopenawa Yanomami (1992), 'Yanomami in peril', *Supranational Monitor*, **13** (9), 1–2.

Davison, A (2004), 'Sustainable technology: beyond fix and fixation', in R White (ed.), *Controversies in Environmental Sociology*, Cambridge: Cambridge University Press, pp. 132–49.

Dawkins, R (1976), *The Selfish Gene*, Oxford: Oxford University Press.

deDuve, C (1995), *Vital Dust: The Origin and Evolution of Life on Earth*, New York: Basic.

Deisingh, A and M Thompson (2004), 'Biosensors and the detection of bacteria', *Canadian Journal of Microbiology*, **50**, 69–77.

Dennett, D (2006), *Breaking the Spell: Religion as a Natural Phenomenon*, New York: Viking.

Deplazes, A (2009), 'Piecing together a puzzle: an exposition of synthetic biology', *EMBO Reports*, **10** (5), 428–32.

Dewey, J (1957), *Human Nature and Conduct*, New York: The Modern Library.

Dietrich, B (1963), *Ethics*, ed. by E Bethge, trans. by N Horton Smith, London: Collins.

Dovers, S (2005), *Environment and Sustainability Policy: Creation, Implementation, Evaluation*, Sydney: Federation Press.

Drexler, KE (1990), *Engines of Creation: The Coming Era of Nanotechnology*, Oxford: Oxford University Press.

Dworkin, R (1998), *Law's Empire*, Oxford: Hart Publishing.

Editorial (2007), 'Tackling global poverty', *Nature Nanotechnology*, **2** (11), 661.

Elechiguerra, JL et al. (2005), 'Interaction of silver nanoparticles with HIV-1', *Journal of Nanobiotechnology*, **3**, 6doi:10.1186/1477-3155-3-6.

European Parliament (2009), 'Report on regulatory aspects of nanomaterials (2008/2208(INI))', Committee on the Environment, Public Health and Food Safety. A6-0255/2009. Rapporteur: Carl Schlyter. 24 April 2009.

Faunce, T (2007), 'Nanotechnology in global medicine and human biosecurity: private interests, policy dilemmas and the calibration of public health law', *Journal of Law, Medicine and Ethics (US)*, **35** (4), 629–42.

Faunce, T (2007), 'Nanotherapeutics: new challenges for safety and cost-effectiveness regulation in Australia', *Medical Journal of Australia*, **186** (4), 189–91.

Faunce, T (2008), 'Toxicological and public good considerations for the regulation of nanomaterial-containing medical products', *Expert Opinion in Drug Safety*, **7** (2), 103–6.

Faunce, T (2011), 'Governing nanotechnology for solar fuels: towards a jurisprudence of global artificial photosynthesis', *Renewable Energy Law and Policy*, **2**, 163–8.

Faunce, T (2011), 'Will international trade law promote or inhibit global artificial photosynthesis?', *Asian Journal of WTO and International Health Law and Policy (AJWH)*, **6**, 313–47.

Faunce, T (2011), 'Global artificial photosynthesis: a scientific and legal introduction', *Journal of Law and Medicine*, **19**, 275–81.

Faunce, T (2012), 'Future perspectives on solar fuels', in T Wydrzynski and W Hillier (eds), *Molecular Solar Fuels*, Cambridge: Royal Society of Chemistry, pp. 506–28.

Faunce, T and K Shats (2007), 'Researching safety and cost-effectiveness in the life cycle of nanomedicine', *J Law Med.*, **15** (1), 128–35.

Faunce, T and A Wattal (2010), 'Nanosilver and global public health: international regulatory issues', *Nanomedicine*, **5** (4), 617–32.

Felming, JR (2007), 'The climate engineers', *Wilson Quarterly*, Spring, 46–60.

Feynman, RP (1992), *The Character of Physical Law*, London: Penguin.

Fortina, P et al. (2005), 'Nanobiotechnology: the promise and reality of new approaches to molecular recognition', *Trends in Biotechnology*, **23** (4), 168–73.

Fox, S (1986), 'Molecular selection and natural selection', *Quarterly Review of Biology*, **61** (3), 375–86.

Freeman, D (1981), *Disturbing the Universe*, London: Pan.

French, D (2000), 'Developing states and international environmental law', *International and Comparative Law Quarterly*, **49**, 35–50.

French, D (2007), 'Managing global change for sustainable development', *International Environmental Agreements*, **7**, 209–11.

Galbraith, K (1984), *The Affluent Society*, 4th edn, London: Penguin.

Goodall, C (2008), *Ten Technologies to Save the Planet*, London: Green Profile.

Graham, G (2008), *Ethics and International Relations*, Oxford: Blackwell.

Green, D (1991), *Writer, Reader, Critic*, Sydney: Primavera Press.

Greene, B (2011), *The Hidden Reality: Parallel Universes and the Deep Laws of the Cosmos*, London: Allen Lane.

Gribbin, J (2009), *In Search of the Multiverse*, London: Penguin.

Gribbin, J (2011), *The Reason Why: The Miracle of Life on Earth*, London: Allen Lane.

Grousset, R (1971), *In the Footsteps of the Buddha*, trans. by JA Underwood, New York: Orion Press.

Gust, D and TA Moore (1989), 'Mimicking photosynthesis', *Science*, **244**, 35–41.

Hamilton, WD (1975), 'Innate social aptitudes of man: an approach from evolutionary genetics', in R Fox (ed.), *Biosocial Anthropology*, New York: Halsted Press, pp. 133–55.

Hansen, J (2009), *Storms of my Grandchildren: The Truth about the Coming Climate Catastrophe and our Last Chance to Save Humanity*, London: Bloomsbury.

Hart, HLA (1962), *The Concept of Law*, London: Oxford University Press.

Hart, SL and MB Milstein (1999), 'Global sustainability and the creative destruction of industries', *Sloan Management Review*, Fall, 23–33.

Hassan, MHA (2007), 'Building capacity in the life sciences in the developing world', *Cell*, **131**, 433–6.

Hawken, P, AB Lovins and LH Lovins (1999), *Natural Capitalism: The Next Industrial Revolution*, London: Earthscan.

Hazen, RM (2006), *Genesis: The Scientific Quest for Life's Origin*, Washington: Joseph Henry Press.

Heehan, P (2008), 'The new global growth path: implications for climate change analysis and policy', *Climate Change*, **91**, 211–31.

Holcombe, RG (1997), 'A theory of the theory of public goods', *Review of Austrian Economics*, **10** (1), 1–22.

Honoré, T (2007), 'The basic norm of a society', in SL Paulson and BL Paulson (eds), *Normativity and Norms*, Oxford: Clarendon Press, pp. 89–112.

Hu, J (2004), 'The role of international law in the development of WTO law', *Journal of International Economic Law*, **7**, 143–57.

Humphreys, S (ed.) (2010), *Human Rights and Climate Change*, Cambridge: Cambridge University Press.

Hurst, JK (2010), 'In pursuit of water oxidation catalysts for solar fuel oxidation', *Science*, **328**, 315–16.

IFOAM Position Paper (2011), *Use of Nanotechnologies and Nanomaterials in Organic Agriculture*, Bonn: IFOAM Head Office.

Invernizzi, N, G Foladori and D Macluran (2008), 'Nanotechnology's controversial role for the south', *Science, Technology and Society*, **13** (1), 123–48.

Jackson, J (2009), *Sovereignty, the WTO, and Changing Fundamentals of International Law*, Cambridge: Cambridge University Press.

Jain, KK (2007), 'The role of nanobiotechnology in drug discovery', *Drug Discovery Today*, **10** (21), 1435–42.

Jianrong, C et al. (2004), 'Nanotechnology and biosensors', *Biotechnology Advances*, **22**, 505–18.

Johnstone, B (2011), *Switching to Solar: What We Can Learn from Germany's Success in Harnessing Clean Energy*, New York: Prometheus.

Kanan, MW and DG Nocera (2008), 'In situ formation of an oxygen-evolving catalyst in neutral water containing phosphate and CO_2', *Science*, **321**, 1072–5.

Kant, I (1986), *Critique of Pure Reason*, trans. by N Kemp Smith, London: Macmillan.

Kant, I (1991), 'Introduction to the metaphysics of morals', in H Reiss (ed.), HB Nisbet (trans.), *Kant: Political Writings*, Cambridge: Cambridge University Press.

Kant, I (2006), 'Introduction to the doctrine of virtue', in I Kant, *Practical Philosophy*, trans. by M Gregor, Cambridge: Cambridge University Press.

Kapczynski, A et al. (2005), 'Addressing global health inequities: an open licensing approach for university innovations', *Berkeley Technol Law J.*, **20** (2), 1031–114.

Karakosta, C, H Doukas and J Psarras (2010), 'Technology transfer through climate change', *Renewable and Sustainable Energy Reviews*, **14**, 1546–7.

Keith, DW (2000), 'Geoengineering the climate: history and prospect', *Annual Review of Energy and Environment*, **25**, 245–84.

Keynes, JM (1931), *Essays in Persuasion*, London: Macmillan.

Kolb, EW and MS Turner (1990), *The Early Universe*, Redwood City, CA: Addison Wesley.

Lahav, N (1999), *Biogenesis: Theories of Life's Origin*, Oxford: Oxford University Press.

Lazcka, O, FJ del Campo and FX Munoz (2007), 'Pathogen detection: a perspective on traditional methods and biosensors', *Biosensors and Bioelectronics*, **22**, 1205–17.

LeDuc, PR et al. (2007), 'Towards an in vivo biologically inspired nanofactory', *Nature Nanotechnology*, **2**, 3–7.

Levskaya A, AA Chevalier et al. (2005), 'Synthetic biology: engineering *Escherichia coli* to see light', *Nature*, **438**, 441–2.

Lewis, NS and DG Nocera (2006), 'Powering the planet: chemical challenges in solar energy utilization', *PNAS*, **103** (43), 15729–35.

Li, N et al. (2003), 'Ultrafine particulate pollutants induce oxidative stress and mitochondrial damage', *Environmental Health Perspectives*, **111** (4), 455–60.

Loncto, J, M Walker and L Foster (2007), 'Nanotechnology in the water industry', *Nanotechnology Law and Business*, **4** (2), 157–9.

Lovelock, J (2006), *The Revenge of Gaia*, London: Penguin.

Luf, G (2007), 'On the transcendental import of Kelsen's basic norm', in SL Paulson and BL Paulson (eds), *Normativity and Norms: Critical Perspectives on Kelsenian Themes*, London: Clarendon Press, pp. 221–34.

Lynch, I et al. (2007), 'The nanoparticle-protein complex as a biological entity: a complex fluids and surface challenge for the 21st century', *Adv in Colloid and Interface Science*, 134–5, 167–74.

Malthus, TR (1960), *On Population*, ed. by G Himmelfarh, London: Random House.

Mannino, S and M Scampicchio (2007), 'Nanotechnology and food quality control', *Veterinary Research Communications*, **31** (Suppl 1), 149–51.

Mannix et al. (2008), 'Nanomagnetic actuation of receptor-mediated signal transduction', *Nature Nanotechnology*, **3**, 36–40.

Marchant, GE and DJ Sylvester (2006), 'Transnational models for regulation of nanotechnology', *The Journal of Law, Medicine & Ethics*, **34** (4), 714–25.

Maskus, KE and JH Reichman (2005), *International Public Goods and Transfer of Technology Under a Globalised Intellectual Property Regime*, Cambridge: Cambridge University Press.

Matthews, HD and K Caldeira (2007), 'Transient climate-carbon simulations of planetary engineering', *Proceedings of the National Academy of Sciences*, **104**, 9949–54.

Matthews, HD, NP Gillett et al. (2009), 'The proportionality of global warming to cumulative carbon emissions', *Nature*, **459**, 829–33.

McKibben, B (1990), *The End of Nature*, New York: Penguin.

Meinshausen, M et al. (2009), 'Greenhouse-gas emission targets for limiting global warming to 2°C', *Nature*, **458**, 1158–62.

Morones, JR, JL Elechiguerra et al. (2005), 'The bactericidal effect of silver nanoparticles', *Nanotechnology*, **16**, 2346–53.

Morse, S (2010), *Sustainability: A Biological Perspective*, Cambridge: Cambridge University Press.

Moses, H III et al. (2005), 'Financial anatomy of biomedical research', *JAMA*, **294** (11), 1333–42.

Mosley, G (2010), *Steady State: Alternative to Endless Economic Growth*, Canterbury: Envirobook.

Murashov, V and J Howard (2008), 'The US must help set international standards for nanotechnology', *Nature Nanotechnology*, **3**, 635–6.

Myers, N (1990), *The Gaia Atlas of Future Worlds: Challenge and Opportunity in an Age of Change*, London: Gaia Books.

Nasu, H and T Faunce (2010), 'Nanotechnology and the international law of weaponry: towards international regulation of nano-weapons', *Journal of Law, Information and Science*, **20**, 21–54.

Nasu, H and T Faunce (2012), 'The proposed ban on certain nanomaterials for electrical and electronic equipment in Europe: global security implications', *European Journal of Law and Technology*, **3** (2), http://ejlt.org//article/view/79.

Nel, A, T Xia et al. (2006), 'Toxic potential of materials at the nanolevel', *Science*, **311** (5761), 622–7.

Nichols-Richardson, SM (2007), 'Nanotechnology: implications for food and nutrition professionals', *Journal of the American Dietetic Association*, **107** (9), 1494–7.

Nijhara, R and K Balakrishnan (2006), 'Bringing nanomedicines to market: regulatory challenges, opportunities and uncertainties', *Nanomedicine*, **2**, 127–36.

Oberdorster, G et al. (2005), 'Nanotoxicology: an emerging discipline evolving from studies of ultrafine particles', *Environ Health Perspect*, **113** (7), 823–39.

OECD (2007), 'Current Developments/Activities on the Safety of Manufactured Nanomaterials'. *Env, Health and Safety Pub. Series on the Safety of Manufactured Nanomaterials No 3.* OECD Environmental Directorate Paris.

O'Neill, K (2009), *The Environment and International Relations*, Cambridge: Cambridge University Press.

Pace, R (2005), 'An integrated artificial photosynthesis model', in A Collings and C Critchley (eds), *Artificial Photosynthesis: From Basic Biology to Industrial Application*, Weinheim: Wiley-VCH Verlag.

Page, EA (2007), *Climate Change, Justice and Future Generations*, Cheltenham, UK and Northampton, MA, USA: Edward Elgar.

Pardo-Guerra, JP and F Aguayo (2005), 'Nanotechnology and the international regime on chemical and biological weapons', *Nanotechnology Law and Business*, **2** (1), 55–61.

Pejcic, B, R de Marco and G Parkinson (2006), 'The role of biosensors in the detection of emerging infectious diseases', *The Analyst*, **131**, 1079–90.

Pittock, B (2009), *Climate Change: The Science, Impacts and Solutions*, London: Earthscan CSIRO.

Plamenatz, J (1963), *Man and Society*, London: Longmans.

Poland, CA et al. (2008), 'Carbon nanotubes introduced into the abdominal cavity of mice show asbestos-like pathogenicity in a pilot study', *Nature Nanotechnology*, Advance Online Publication 2008, 20 May, 1–6.

Presting, H and U Konig (2003), 'Future nanotechnology developments for automotive applications', *Materials Science & Engineering*, **23**, 737–41.

Resii, H (ed.), Nisbet HB (trans.) (1991), 'Conjectures on the beginning of human history', in *Kant: Political Writings*, 2nd edn, Cambridge: Cambridge University Press, pp. 231–2.

Resnik, DB and SS Tinkle (2007), 'Ethics in nanomedicine', *Nanomedicine*, **2** (3), 345–50.

Ridley, M (1997), *The Origins of Virtue, Human Instincts and the Evolution of Cooperation*, New York: Viking.

Robock, A (2000), 'Volcanic eruptions and climate', *Reviews of Geophysics*, **38**, 191–219.

Royal Society and Royal Academy of Engineering (RS/RAE) (2004), *Nanoscience and Nanotechnologies.* Policy Doc 19/04.

Russell, B (1930), *The Conquest of Happiness*, London: Allen & Unwin.

Salicrup, LA and L Fedorkova (2006), 'Challenges and opportunities for enhancing biotechnology and technology transfer in developing countries', *Biotechnol Adv.*, **24** (1), 69–79.

Sanderson, K (2008), 'The photon trap', *Nature*, **452**, 400–402.

Sandin, P (1999), 'Dimensions of the precautionary principle', *Human & Ecological Risk Assessment*, **5**, 889–907.

Saul, JR (1995), *The Doubter's Companion: A Dictionary of Aggressive Common Sense*, London: Penguin.

Saunders, N (2002), *Divine Action and Modern Science*, Cambridge: Cambridge University Press.

Savage, N and M Diallo (2005), 'Nanomaterials and water purification: opportunities and challenges', *Journal of Nanoparticle Research*, **7**, 331–42.

Schelling, TC (1996), 'The economic diplomacy of geoengineering', *Climatic Change*, **33**, 291–302.

Schneider, ED and JJ Kay (1994), 'Life as a manifestation of the second law of thermodynamics', *Mathematical and Computer Modeling*, **19** (6–8), 25–48.

Schneider, ED and D Sagan (2005), *Into the Cool: Energy Flow, Thermodynamics and Life*, Chicago: University of Chicago Press.

Schneider, S and W Broecker (2007), 'Geoengineering may be risky but we need to explore it', *New Scientist*, **195**, 44–5.

Schneider, SH, JR Miller, E Crist and PJ Boston (eds) (2004), *Scientists Debate Gaia: The Next Century*, Cambridge, MA: MIT Press.

Scholte, JA (2000), *Globalization: A Critical Introduction*, London: Macmillan.

Schopenhauer, A (1948), *The World as Will and Idea*, Vol. 1, London: Routledge & Kegan Paul.

Schroedinger, E (1944), *What Is Life?* Cambridge: Cambridge University Press.

Schulte, P, C Geraci et al. (2008), 'Occupational risk management of engineered nanoparticles', *Journal of Occupational and Environmental Hygiene*, **5**, 239–49.

Schumacher, EF (1974), *Small Is Beautiful: Economics As If People Mattered*, London: Abacus.

Seckbach, J, J Chela-Flores, T Owen and F Raulin (eds) (2004), *Life in the Universe: From the Miller Experiment to the Search for Life on Other Worlds*, Dordrecht: Kluwer.

Segrè, G (2008), *Faust in Copenhagen: A Struggle for the Soul of Physics and the Birth of the Nuclear Age*, London: Pimlico.

Sen, A (2010), *The Idea of Justice*, London: Penguin.

Service, RF (2009), 'Hydrogen cars: fad or the future?' *Science*, **324**, 1257–9.

Shaunak, S and S Brocchini (2007), 'Entrepreneurial experiences', *Nat. Rev Drug Discovery*, **6**, 499.

Shine, KP et al. (2007), 'Comparing the climate effect of emissions of short- and long-lived climate agents', *Phil Trans R Soc*, **365**, 1903–14.

Shing-Tung, Y and S Nadis (2010), *The Shape of Inner Space: String Theory and the Geometry of the Universe's Hidden Dimensions*, New York: Basic Books.

Shrivastava, P (1995), 'The role of corporations in achieving ecological sustainability', *Academy of Management Review*, **20** (49), 936–60.

Silver, S, LT Phung and G Silver (2006), 'Silver as biocides in burn and wound dressings and bacterial resistance to silver compounds', *Journal of Industrial Microbiology and Biotechnology*, **33**, 627–34.

Singer, P (2002), *One World: The Ethics of Globalisation*, Melbourne: Text Publishing.

Smagadi, A (2006), 'Analysis of the Objectives of the Convention on Biological Diversity', *Columbia Journal of Environmental Law*, **31**, 243–56.

Sober, E and DS Wilson (1998), *Unto Others: The Evolution and Psychology of Unselfish Behavior*, Cambridge, MA: Harvard University Press.

Sobolski, GK, JH Barton and EJ Emanuel (2005), 'Technology licensing', *JAMA*, **294** (24), 3137–40.

Sridhar, D, S Khagram and T Pang (2008), 'Are existing governance structures equipped to deal with today's global health challenges? Towards systematic coherence in scaling up', *Global Health Governance*, **2** (2), 1–25.

Stager, C (2011), *Deep Future: The Next 100,000 Years on Earth*, Melbourne: Scribe.

Stead, JG and WE Stead (2009), *Management for a Small Planet*, 3rd edn, Sheffield: Greenleaf Publishing.

Stone, C (1985), 'Should trees have standing? Revisited: how far will law and morals reach? A pluralist perspective', *Southern California Law Review*, **59**, 1–27.

Sunstein, CR (2005), *Laws of Fear: Beyond the Precautionary Principle*, Cambridge: Cambridge University Press.

The Royal Society (2007), *Nanoscience and Nanotechnologies: Opportunities and Uncertainties.* RS Policy Doc 19/04 July 2004 WPNM policy doc 13/07 April 2007.

Trop, M et al. (2006), 'Silver coated dressing Acticoat caused raised liver enzymes and argyria-like symptoms in burn patient', *Journal of Trauma-Injury Infection and Critical Care*, **60** (3), 648–52.

Tumpey, TM et al. (2005), 'Characterisation of the reconstructed 1918 Spanish influenza pandemic virus', *Science*, **310**, 77–80.

Turner, GA (2008), 'A comparison of the limits to growth with 30 years of reality', *Global Environmental Change*, **18** (3), 397–411.

Twining, W (2009), *General Jurisprudence: Understanding Law from a Global Perspective*, Cambridge: Cambridge University Press.

United Nations (2009), -/CP.15 Conference of the Parties, Fifteenth session Copenhagen, 7–18 December 2009 FCCC/CP/2009/L.7.

US FDA (2007), *Regulation of Nanotechnology*. United States Food and Drug Administration. http://www.fda.gov/nanotechnology/regulation.html, FDA 19 December 2007.

US NIOSH (2007), *Evaluation of Health Hazard and Recommendations for Occupational Exposure to Titanium Dioxide*. CDC. http://www.cdc.gov/niosh/review/public/TIo2/pdfs/TIO2Draft.pdf, 1 November 2007.

Voigt, C (2009), *Sustainable Development as a Principle of International Law*, Leiden: Martinus Nijhoff.

Wardak, A and ME Gorman (2006), 'Using trading and life cycle analysis to understand nanotechnology regulation', *Journal of Law, Medicine and Ethics*, Winter, 695–713.

Warheit, KL et al. (2004), 'Comparative pulmonary toxicity assessment of single-walled carbon nanotubes in rats', *Toxicological Sciences*, 77, 117–25.

Weinberg, S (1977), *The First Three Minutes*, New York: Basic Books.

Wijnhoven, SWP et al. (2009), 'Nanosilver: a review of available data and knowledge gaps in human and environmental risk assessment', *Nanotoxicology*, 3 (2), 109–38.

Wilczek, F (2008), *The Lightness of Being: Mass, Ether and the Unification of Forces*, New York: Basic Books.

Wilson, EO (1978), *On Human Nature*, Cambridge, MA: Harvard University Press.

Yohe, GW, R Tol, R Richels and G Blanford (2009), 'Climate change', in B Lomborg (ed.), *Global Crises, Global Solutions*, Cambridge: Cambridge University Press.

Index